Sociological Research

A Philosophy of Science Perspective

Sponsored by the Sociological Institute for Community Studies
Bar-Ilan University, Israel

Sociological Research

A Philosophy of Science Perspective

by

Ernest Krausz

VAN GORCUM 1986
Assen/Maastricht, The Netherlands
Wolfeboro, New Hampshire 03894-2069, U.S.A.

Library of Congress Cataloging-in-Publication Data

Krausz, Ernest.
Sociological research.

At head of title: Sponsored by the Sociological Institute for Community Studies, Bar-Ilan University, Israel.
Bibliography: p.
Includes indexes.
1. Sociology-Methodology. 2. Sociology-Research.
I. Title.
HM24.K665 1986 301'.072 86-19433

ISBN 90-232-2207-5 (pbk.)

Printed in the Netherlands by Van Gorcum, Assen

Table of Contents

Acknowledgments

First and foremost I wish to express my gratitude to the Sociologica Institute for Community Studies at Bar-Ilan University for sponsoring the sociological publication of this book.

My grateful thanks are due to many colleagues who have given me valuable advice, and to several universities in England for providing me with excellent facilities during sabbaticals and shorter visits: Peter Collison, Kathleen Bell, Norman Dennis, and Elizabeth Gittus at the University of Newcastle; Donald G. MacRae, Ernest Gellner, David Martin, and Percy S. Cohen at the London School of Economics; Julius Gould at the University of Nottingham; Stanislav Andreski, and Christie Davies at the University of Reading; David Lane at the University of Birmingham; Stephen H. Miller, and Ardon Lyon at the City University, London. Last but not least I am grateful for the constant encouragement and valuable advice I received from colleagues at Bar-Ilan University: Max Jammer, Bernard Lazerwitz, Eliezer Goldman, Gitta Tulea, and Yakov Reuveni.

I benefited a great deal from the sessions I attended at the 7th International Congress of Logic, Methodology and Philosophy of Science which took place in Salzburg, Austria in July 1983. For that I thank my hosts at the University of Salzburg, in particular Paul Weingartner and Otto Neumaier.

Finally my thanks are due to Susie Grossman for the expert typing of the manuscript.

E.K.

Chapter 1

Introduction

The purpose of this work is to set forth the principles underlying sociological research. It will be helpful in this respect to make a distinction between 'methods' in the plural, referring to the detailed techniques of work, and 'method' in the singular, referring to general principles (Ravetz, 1971). Any reflective social scientist, who is not so absorbed in specialized research as to lose sight of fundamentals, must concern himself with the philosophical foundations of his scientific work (Pap, 1962). But it is not our intention to develop an abstract and self-contained methodological discourse which does not relate to the actual work done in the discipline of sociology.

If Poincaré's remark that "while physicists had a subject-matter, sociologists were engaged almost entirely in considering their methods" (Cohen, 1953) had once a true ring about it, this has not been the case for a long while. For many decades sociologists have accumulated rich material on the specialized subject-matter of man's collective and interdependent life in society. This material about the social life-world has been gathered and analyzed by means of a variety of techniques, such as sample surveys and statistical analysis, comparative case studies, or the sifting of the contents of documentary material. Often sociologists simply took it for granted or decided after brief consideration that one or other of these techniques was suitable for the particular research topic and situation they were confronting. In recent years, however, sociologists have become more aware of the need to assess the principles which "are implicit in the details of research practice" (Pawson, 1978; Banks, 1979).

In focusing on such principles we shall assume that a detailed account of the techniques of research is not required for our purposes. Thus, our interest is not in how one can construct indices based on empirical indicators, but rather in how the sociologist's conceptualizations link up with such indicators. Again, it is not the minutiae of survey or experimental designs that will arrest our attention, but the time factor and its effects through history and maturation. In this sense our distinction between the techniques of research and their underlying philosophical principles leads us to the acceptance of the definition of 'methodology' as "the study of the principles of scientific procedures; that is to say, its basic premises and philosophical foundations"

(Machlup, 1978).

The concern with methodology among scientists stems from the natural expectation within any community of scholars that there should be some statement, if not a complete scheme, of not only the source of the research material and the manner in which the account of life and phenomena has been arrived at, but also the logic upon which that account rests. This short book is one such statement. Its structure may be presented by means of five major questions that we pose: 1. What is the significance of 'variables' in the language and practice of social research? 2. Is the principle of causality applicable to social phenomena? 3. What are the time constraints in sociological research? 4. What is social reality? and 5. In what way can testability be achieved in the study of society?

The selection of these themes cannot be justified entirely on logical grounds, although the logical connections between them will become clear as we develop them. And it is undeniable that other important questions could have been asked within our methodological framework. For instance, one could have asked: what are the principles upon which the selection of sociological data is based and how does one ensure the adequacy and relevance of such data for the problem investigated; or what are the effects of the judgmental and intuitive aspects of problem formulation and data analysis on scientific results. We feel that although we do not deal with such other questions, this does not greatly diminish the value of our discussion. The focus on our selected topics is strengthened by the unifying threads which run through the work. These are: (a) the belief in the unity of science, a belief that has been bolstered by the recognition that basic uncertainties exist in all the sciences, including physics; (b) the basically Popperian notion that testability is the crucial scientific criterion, and the view we hold that this principle can be followed in social research; and (c) that since in the sociological domain laboratory experiments are not feasible, the principle of testability may best be upheld in the applied field; that is, through the natural laboratory of the social world, which makes a science of society possible.

Let us now provide the reader with the rationale that binds together the questions we posed above. We can do this by outlining the examination to which these questions will be subjected in the chapters that follow.

Sociological research, as all scientific research, has a dual nature: abstraction and operationalization. The clearest expression of this duality is in the use of concepts and indicators. The basic question that arises is how one can ensure a true correspondence between our abstract conceptualizations and the indicator-type measurements we carry out in research. This is where variables come in. As mediating constructs they provide the bridging between our scientific concepts and the real-world phenomena. Variables can do this job since they combine both conceptual and empirical elements.

Now, the central aim of research is explanation based on empirical evidence. To achieve this we need some methodological framework which relates to reality, i.e., some logical arrangement of reality which we can understand. Variables, such as social class or ethnicity, cannot be considered haphazardly, and simply describing and enumerating them is not sufficient. Hence we use models and theoretical propositions. Our aim in using such patterning is to create *order* and, thereby, to uncover the *causal relationships* between phenomena. This leads us to the question whether we can achieve true causal explanation in social research. That is to say, is the principle of causality applicable to social phenomena? And even if we can assume that it is, how can we establish causal connections between variables with any degree of confidence? And as a corollary, we can pose the further question whether sociological research can result in lawlike propositions.

These questions arise because of the problem in social research of clearly demonstrating causal relationships. The difficulty is due in large measure to our inability to carry out laboratory experiments. Such experimentation is crucial in the natural sciences, because it facilitates the manipulation and control of the variables involved in the model set up, and because it can neutralize extraneous factors. One such factor is time. It is the inability of the sociologist to neutralize time in the social world, more than his difficulties of controlling and manipulating variables, that prevent him from carrying out rigorous experimental research. Furthermore, the time factor affects even his non-experimental research efforts. This is why we proceed to consider the time constraints on social research, and also provide a brief and basic introduction to the nature of time and the philosophical conceptions of this element.

The fact that the sociologist does his research in non-laboratory conditions raises the question of how he can test his theoretical propositions. Here we develop the argument that the sociologist has no option other than using social reality as his testing ground. The sociologist must, therefore, face what is, perhaps, *the* basic question in social research: does he apprehend and *re*construct a given social reality? or, does he simply create his own reality? The latter possibility does lurk in the background, and it could lead to an interminable 'realist versus instrumentalist' dispute. This brings us back full circle to the question of how testability can at all be attained. We find an answer in the 'applied science' argument.

Applied sociology is concerned with diagnosing and solving social problems and is close to policy formation and practical decision. As such it requires high quality prediction and forecasting, requirements which are particularly difficult to attain in view of the various constraints under which the sociologist undertakes his research, the time constraint itself being a heavy hindrance in the applied context. Nevertheless, the adherence to a more basic sociological science, unconcerned with the question of testability for its universal, abstractly conditional, theoretical statements does not offer the hope of scientific advancement in the study of society. As Piotr Sztompka says: "Sociology is also seen as directly relevant for social practice, and social practice as providing ultimate corroboration of sociological knowledge" (Sztompka, 1982). If Sztompka sounds over-optimistic in hoping for ultimate corroboration of sociological knowledge, it is certainly true that, despite the difficulties, applied research does provide the testing ground for sociological theories.

We conclude our introductory remarks by stressing that we shall develop our sociological methodology within an indeterminist scientific framework. The two pivots of such a sociology are: bounded rationality, and probabilistic explanation and prediction. As such, we believe that the social world is scientifically researchable, although the results of sociological research will provide us only with incomplete knowledge. However, in the efforts to improve our understanding of the workings of society, what matters is not so much how perfect such knowledge and understanding may be, but the recognition that, given certain conditions, a scientific approach is feasible. This is, in essence, what we aim to show in the pages that follow.

Bibliography

Cohen, Morris, R., 1953: *Reason and Nature – An Essay on the Meaning of Scientific Method*, The Free Press, Glencoe, 2nd edition, p. 350.

Banks, J.A., 1979: "Sociological Theories, Methods and Research Techniques", *The Sociological Review*, Vol. 27, No. 3, 1979, p. 563.

Krausz, Ernest, 1984: "In Defence of Methodological Collectivism", in O. Neumaier, ed., *Mind, Language and Society*, Conceptus Studien 2, VWGO, Vienna, pp. 145-157.

Krausz, Ernest, 1985: "The Principle of Causality and Its Application to Social Phenomena", *Methodology and Science*, Vol. 18, No. 3, 1985, pp. 225-240.

Machlup, Fritz, 1978: *Methodology of Economics and Other Social Sciences*, Academic Press, New York, p. 54.

Pap, Arthur, 1962: *An Introduction to the Philosophy of Science*, The Free Press, Glencoe.

Pawson, Ray, 1978: "Empiricist Explanatory Strategies: The Case of Causal Modelling", *The Sociological Review*, Vol. 26, No. 3, 1978, p. 614.

Pawson, Ray, and Tilley, Nicholas, 1982: "Monstrous Thoughts: Weaknesses in the Strong Programme of the Sociology of Science", Occasional Papers, No. 14, University of Leeds, December, 1982.

Popper, Karl, R., 1983: *Realism and the Aim of Science*, From the „Postscript to the Logic of Scientific Discovery", ed., by W.W. Bartley III, Hutchinson, London.

Ravetz, Jerome R., 1971: *Scientific Knowledge and Its Social Problems*, Clarendon Press, Oxford, pp. 146, 149.

Smart, J.J.C., 1982: "Difficulties for Realism in the Philosophy of Science", in *Logic, Methodology and Philosophy of Science*, ed., by L.J. Cohen, et al., North Holland Publishing Company and Polish Scientific Publishers, Amsterdam and Warsaw, pp. 363f.

Sztompka, Piotr, 1982: "Metatheoretical Dilemmas of the Social Sciences – the Case of Sociology", in *Logic, Methodology and Philosophy of Science*, ed., by L.J. Cohen et al., North Holland Publishing Company and Polish Scientific Publishers, Amsterdam and Warsaw, p. 387.

Chapter 2

Concepts, Variables, and Indicators

Robert K. Merton in his classic statement on the bearing of empirical research on sociological theory has not only emphasized the close and reciprocal relationship between sociological theory and empirical research, but has also recognized the operationalist needs of sociological research (Merton, 1957). It is not our purpose, however, to reaffirm the realist approach. Rather, our aim is to consider the problems which confront those who relate their theories and research to any assumed 'real' world. It is, therefore, on the assumption that the world we live in has some empirically founded basis for inter-subjective agreement that we proceed to examine the modes of operationalization of concepts, the function of mediating variables, and the use of indicators in a scientific process which aims at apprehending and understanding social reality.

Variables as mediating concepts

We start our discussion with the basic premise that there is continuity and interaction between the everyday world and social science, and our specific interest is to show how such interaction is effected by the identification of variables and their concrete operationalized measurement by means of indicators. The research procedure is also characterized, though, by the need to select, abstract and categorize certain aspects of reality in order to facilitate a clearer vision of that reality and thus enable the scientist to carry out orderly analysis of the factual world. In this respect scientific constructs i.e. concepts, play an important role since they enable us to disentangle scientific communication from everyday language. This is most important in an area in which the reciprocal influences of the two are likely to occur frequently and as a result of which the scientific endeavour may be adversely affected.

We must make it clear, however, that by disentangling the two languages we do not necessarily run the risk, as some fear (Cicourel, 1964), of severing our scientific pursuit from the real phenomena of the social world we study, thus distorting, through reification and generalization, our understanding of the real world. We avoid this danger by identifying and using yet another type of construct, which we shall call 'mediating constructs'. These are the 'variables' we use in operationalizing our research, both pure and applied.

Let us illustrate the way in which variables mediate between the everyday world we study and the scientific enterprise we use to carry out such study. The very essence of identifying variables is the ability to recognize the measurable, testable and, where possible, manipulable areas of the real phenomena of the social world. The identification of variables will be instrumental in describing and explaining these phenomena as accurately as possible, both in the objective sense, that is in terms of their external aspects, and in terms of their meanings, that is in the subjective sense. At the same time these variables will be constructed and used on the guidelines laid down by our scientific conceptualizations and thus provide consistent and coherent descriptions leading to scientific analysis and explanation. Our idea of variables as *mediating concepts* is presented below by means of two diagrams:

DIAGRAM 1

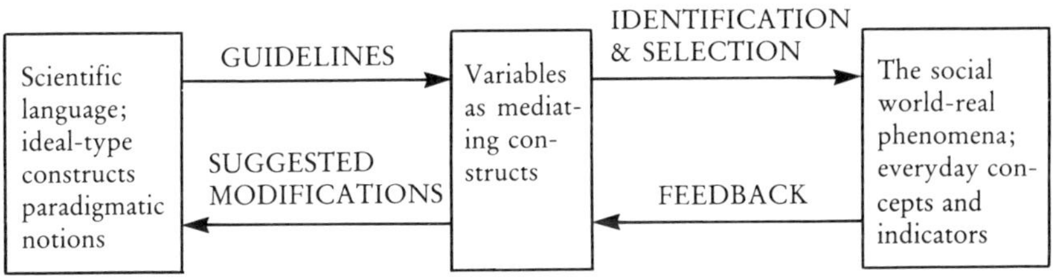

DIAGRAM 2

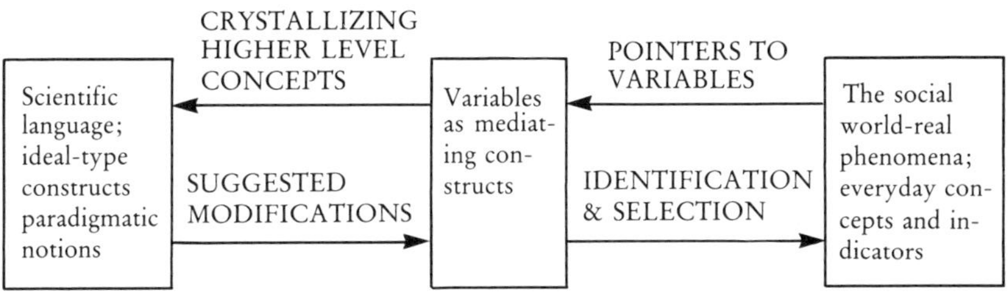

We must imagine these two diagrams as being superimposed on each other in order to gain a feel of the reciprocal processes continuously taking place, whereby scientific constructs and empirical indicators mutually influence one another. As our schemata suggest, there is an ongoing process of identification and selection of operationalizable variables from among everyday social phenomena. The identification of these variables is based on both the theoretical guidelines scientifically conceptualized and the pointers provided by the real-world situation. The process is a complex one, for the very operationalization of the variables often provides the kind of feedback which suggests the need for modifications and the wider reformulations of our scientific concepts. This means that the scientific guidelines are not so rigidly set as not to allow changes in the selection of variables made necessary by changes in the real-world phenomena. On the contrary, our scientific concepts must be sensitive enough to monitor such changes, by means of showing up logical inconsistencies or weaknesses in our ability to identify through them meaningful variables. On the other hand, by stressing the need to identify operationalizable variables we do not adopt a radical operationalist stance, since the latter would dismiss theoretical constructs except as lables for operationally defined events (Krausz & Miller, 1974). As we shall see below, operational indicators unconnected with scientific constructs are likely to be of little value in social research. But we must reiterate that this is not to imply that the indicators themselves do not influence the very construction of our scientific concepts. The social scientist, as any other scientist, produces scientific concepts so as to order and understand a very complex real-world situation, but equally the latter will influence him in the very formulation of his scientific concepts.

In connection with identifying and operationalizing variables it is generally accepted that theoretical concepts, that is what we have called scientific constructs, must be 'ultimately expressible' in terms of observational concepts. According to Peter Abell (1971) the standard practice in Sociology is to operationalize in the following manner: theoretical concepts may at first be expressed in terms of *other* theoretical concepts, as in the case of 'alienation' which is expressed in terms of 'meaninglessness', 'normlessness', 'powerlessness' and 'isolation' (Blauner, 1964); but *ultimately* the latter must be operationalized through observational concepts. The procedure, Abell tells us, is to convert theoretical concepts into theoretical variables, which in turn are operationalized by means of selecting indicators from a set of observational items (Abell, 1971). The indicators themselves are conceptualized facets analytically related to our variables. Thus, our indicator may be education, our variable social class, and our theoretical construct social stratification. When we are measuring the indicator, in our case education, we are measuring some knowledge we have concerning the education received, as for instance the number of years of education, rather than education itself. The analogy here is with Heisenberg's description of atomic physics where the scientist, whose knowledge is always incomplete, does not really or ultimately deal with the elementary particles themselves but with his knowledge of them (Heisenberg, 1955).

Our focus in this section on the identification of variables is crucial for a number of reasons. First, "it is as variables that our concepts will eventually appear in propositions and models" (Abell, 1971). Concepts and models are basic elements in our theoretical approach in investigating real-life situations. The models themselves, in which a number of variables are linked, are isomorphic with reality, not true or full representations of it but reified schemes; they are simplified versions which enable us to grasp the fundamentals of a problem or a situation without having to deal with it in all its complexity from the outset. The variables play, therefore, a key role in our understanding of a problem or a social situation. Second, as our scheme above shows, we regard variables as mediating constructs, bridging our scientific conceptions and our real social phenomena. Furthermore, they provide the most convenient spot where the researcher can cut into or prize open the circular scientific process of the research procedure. This is of particular relevance to applied sociology where the additional requirement of teasing out "variables of potentially useful knowledge" must be invoked (Lazarsfeld, 1968).

In order to meet the latter requirement in applied fields, where one must obtain knowledge that leads to the possibility of action, the variables that will produce such useful knowledge will be those whose empirical indicators are 'engineerable'. Several criteria have been proposed for obtaining the useful knowledge through the selection of variables; chief amongst them that the indicators be *identifiable*, *accessible* and *manipulable* (Lazarsfeld, 1968). Clearly some variables, such as early childhood experience, will hardly meet such criteria. Others, such as social class, will meet them partly in that the indicators of occupation, education, or housing may be identifiable and accessible although less manipulable, whilst some indicators such as ideological commitment would be more difficult to identify. Coming nearer to the ideal variable for applied purposes we could exemplify with the concept of 'educational streaming' in the formal school situation which may suggest the use of form groups as variables whose indicators of size, sex and age composition are identifiable, usually accessible and in principle manipulable.

A rigid insistence on adhering to the above criteria poses, however, a serious dilemma in applied research. Whilst it might be true that we may never be sure of having included all the relevant variables and all their indicator facets in a causal model, except where repeated testing reveals the predictive accuracy of such a model, an insistence on selecting only manipulable indicators would *ipso facto* reduce the validity of our research, since only *some* of the variables or *some* of their indicators would be measured and probed. To overcome this problematic nature of a research strategy which tries "to identify variables representing conditions the policy-maker can manipulate to achieve his goals". Lazarsfeld and Reitz suggest that non-manipulability does not necessarily force the exclusion of such variables from applied research. For although "age cannot be manipulated by the policy-maker . . . age differences in observed behaviour can lead to useful conclusions" (Lazarsfeld & Reitz, 1975). One may say that in such cases *indirect manipulation* of the variable does occur. By observing and comparing behaviour, relevant to the problem investigated, in different age-groups, the variable *age* becomes 'manipulable' in the research process and will influence the conclusions drawn.

Indicator selection and measurement

Given that we have identified our variables by means of the reciprocal interplay between scientific concepts and real-world phenomena, and by applying suitable models to the problem investigated, certain decisions will have to be taken regarding the selection of indicators and particularly in so far as their measurement is concerned. Firstly, we may note that indicators accrue in the social sciences in a great variety of ways. Lazarsfeld talks, for instance, about the possibility of 'thinking up' indicators usually based on the principle of a dimensional analysis which produces good leads. Thus, the variable 'integration' would provide leads for indicators by probing such dimensions as the amount of communication between people, existing prejudices between groups, specialization and interdependence, and so on (Lazarsfeld, 1973). Albert Biderman, on the other hand, describes how indicators evolve through a complicated and imperfect procedure for determining priorities regarding specific items to be included for particular purposes. The inadequacy of existing data could be one factor in adding new items to existing ones. Important political issues may thrust some dimensions to the fore. The interest or articulateness of a bureaucrat or academician could be the deciding factor in other instances. Pressures and balances within an organization and influences from outside groups could also weigh in determining what indicators will be available. Similarly, the skill or ineptitude of social scientists, the prejudices of lay authorities, and the opinions of the public could all influence the selection of social indicators (Bauer, 1966).

The second point we must consider is whether our indicators are to be 'tailor-made' for the specific investigation before us, or whether we shall use, adapt and apply existing indicators and utilize the existing data gathered for them. Such secondary sources are either the regular gathering of data around key social indicators as instanced by the publication *Social Trends* produced at the Central Statistical Office in Britain since 1970, or the kind of specific projects undertaken in the United States in the 1960s for detecting, through a wide range of social indicators, the second-order consequences of the NASA space exploration programme in order to anticipate and control the social impact of that programme (Bauer, 1966). But since such secondary sources of social indicators tend to leave gaps, which become apparent when we are faced with a new problem, the sociologist will normally have to track down additional indicators even where he can utilize some of those already known and established.

Thirdly, there are the problems of measurement and inference. So far as measurement is concerned the problem is a lack of homogeneity in the measurability of indicators and the consequent difficulties of standardizing the measurements applied to different indicators. Thus, income and number of years of formal education, as indicators of social class, can be measured at the ratio level, but occupation, another indicator of social class, can be measured only at the ordinal level, that is to say by ranking or ordering occupations according to higher or lower status. There is the possibility of levelling down in order to bring about homogeneity in our measurement. This can be done by simplifying to a lower level the measurement for the indicators which we could have measured at a higher level. A more dubious possibility is to assume higher level type measurement for indicators which in effect lend themselves for measurement at a lower level. As Abell says, since we tend to condense our indicators into indices, our strategy will be determined partly by the level we are aiming for with regard to the index (Abell, 1971), but ultimately, it would seem to us, by the lowest common denominator of indicator measurability.

As we have just mentioned, there is a tendency to recombine indicators into indices which gives rise in particular to the problem of heterogeneity. But index formation is practised because of the difficulty of operating separately with numerous indicators and the dimensions they represent (Lazarsfeld, 1973); or because "the investigator has a number of measures or indicators of what he considers to be the same underlying theoretical variable", and hence when "all such measures are highly inter-correlated, he may merely select the one measure that he considers most appropriate, or he may combine these indicators into a composite index" (Blalock, 1961). On the other hand, just as often we encounter *fractional measurement* because of a lack of coincidence between the underlying concept of a variable and the inferences made with regard to its measurement by means of indicators operationally defined. As Bernard Cazes stresses, this is due to the multiple dimensions of concepts used in the social sciences. It is often, therefore, the case that "a single indicator is wrongly thought sufficient to measure a concept, whereas other indicators are essential" (Shonfield and Shaw, 1972). Thus, measuring the trend in crime by means of an indicator such as number of police arrests would leave out undetected crime for which no indicator is available. Similarly, an overall index of criminality may mask the relative trends of the different indicators combined, unless a very high degree of inter-correlation is found constantly. And 'half-dimensional measurement' is further exemplified by variables such as 'poverty' where comparability and measurement are bedevilled by the qualitative aspects that are bound to figure in our conceptualization of this variable.

One could, of course, deliberately avoid the problem of fractional measurement by what has been called 'concept reduction' (Etzioni and Lehman, 1967). This is the practice of reducing a concept, by definition, to its measurable elements, as in the case of 'inborn intellectual capacity' which is equated with the scores obtained in intelligence tests. This is a defective solution, since it evades rather than avoids the problem of fractional measurement by discouraging the search for additional indicators and better measurement (Shonfield and Shaw, 1972).

But the opposite could also happen: our indicator may measure more than the particular phenomenon our concept requires that should be measured (Zetterberg, 1965). For instance some indicator of prejudice, such as exclusion from a club, may measure both class and ethnic factors, when in fact we require distinct indicator measurement of ethnic prejudice only, if the latter is the concept we utilize. This example also suggests the possibility of the overlapping of two distinct concepts as reflected in a single indicator (Forcese and Richer, 1973). Alternatively we have the frequent situation, as discussed above, where we test our abstract concepts by means of multiple indicators. The use of multiple indicators is called for particularly when the investigator cannot obtain or defend *a single, unambiguous, direct, operational definition* for his theoretical concept. However, the use of multiple indicators is not necessarily a drawback, for it maximizes his information, thus increasing the empirical variability of his measures; it permits him a more detailed and precise analysis; and it provides a fuller representation of the concept (Curtis and Jackson, 1962; Warren et. al. 1950). Yet difficulties arise where different types of indicators do not show clear-cut relationships and even provide evidence of contradictory trends, although they are supposed to tap information concerning the same global concept. Such a concept might be 'wellbeing' or 'quality of life'. Major research projects in such policy oriented areas have proposed the kind of conceptual models which show the multidimensionality of the concepts and the relationships between subjective and objective indicators. (Andrews and Withey, 1976; Campbell, Converse and Rodgers, 1976, Allardt, 1972).

Rules of correspondence

Our discussion is clearly leading to the issue that methodologists and philosophers of science have variously termed as the problem of *slippage* between indicator and concept, or more generally as the need to apply *rules of correspondence* between abstract conceptions and concrete events (Costner, 1971). Therefore, the problematic nature of rules of correspondence between the conceptual and the empirical may now be examined. To do so we must first understand the process of conceptualization. In its general sense conceptualization is germane to the entire learning process and is carried out by means of the formation of mental images abstracted from percepts. Our more specific concern is scientific or disciplined conceptualization, which is supposed to reach a higher level of precision.

Through this process of conceptualization we sort out and clarify our perceptions of all the things that appear to us to be real; all that we deem to exist, whether physical objects and occurrences, social entities and processes, or psychological states. By abstraction we attempt to lay down some degree of permanency for our perceptions by arriving at certain conceptual conventions, if not at fully agreed-upon terminological definitions. This is not only a convenient way of ordering and attempting to understand the world around us; it is the only way of extending commonsense knowledge into scientific knowledge. The problem with conceptualization is that the degree of abstraction necessary depends on the nature of the phenomenon dealt with and increases especially with the degree of its complexity. Construing the problem somewhat differently, one can say that the degree of abstraction necessary is inversely related to the ease with which the empirical referents can be identified. Thus, the demographic concept of 'fertility' is clearly related to the number of births per female of child-bearing age within a given population. There is not the same ease of linking the empirical referents of a sociological concept like 'social status'.

Let us consider the latter from the point of view of scientific concept formation. There appears to be a two-step procedure involved: *meaning analysis* and *empirical analysis*. In the first step we "attempt to cull out the most basic implicit or explicit assumptions inherent in the various meanings that have been attributed to the concept" (Dumont and Wilson, 1967). Thus, we shall come across the idea that 'social status' is related to occupational prestige, education, income, plus other factors, all of which when taken together produce 'social status', which somehow is more than the sum of its component factors. In the conceptualization of status, however, Max Weber was influential. Social status in his scheme is one of the three major dimensions of social stratification: economic class, status, and power, and is defined as personal honour and prestige. But status was designated by Ralph Linton as a social position occupied by an individual in a social system, such positions being, for instance, infant or married person. These various meanings are often interrelated, yet there is no final consensus among sociologists regarding the exact meaning of 'social status'.

Despite the usual lack of consensus, let us suggest, with Dumont and Wilson, that in this case the customary usage points to two basic assumptions, namely that social status is 'a phenomenon of human perception' referring to honour and prestige, and that it is an 'orderable attribute', i.e. that individuals can be ranked in some hierarchical order for comparative purposes. This will lead to a clarification of the way the defining elements of the concept are used in statements; that is to say, we shall be able to shed light on the type of concept we are dealing with from the contextual aspect. Thus, we can say what syntactical combination is possible for this concept: we shall not make a quantitative statement as to the number of units of social status a particular individual has, but rather state ordinally that individual X has higher status then individual Y.

We are now ready for step-two, that is empirical analysis, which "refers to that process whereby the basic assumptions which have been brought to light as a result of the meaning analysis are submitted to direct empirical test" (Dumont and Wilson, 1967). In doing this we encounter the difficulty that in comparison with a physical property such as magnetism which is not itself observable, but its experiential data are: e.g. iron objects placed in the proximity of a magnetic field move towards it, in the case of social status the defining experiential elements such as 'honour' or 'prestige' are *themselves* not observable. Attempts will be made, therefore, to objectify the concept by means of a host of empirical indicators, such as income, style of life, place of residence, etc., combining them into indices, which take account of the interrelationship between such indicators. The real test is in trying to relate such indices to our concept 'social status', for the purpose of which we have to invoke 'bridging principles'. We have seen above that the bridging is accomplished by identifying the relevant variables. Thus, the concept 'social status' may be translated into variable language by referring to, say, 'status situation'. The latter is partly theoretical and partly empirical, as all bridging statements are; that is to say, the status of an individual or a group is at least partly observable through the status situation. It is this linchpin that constitutes the application of rules of correspondence, that is making sure that our concept in fact deals with the reality we wish to investigate. Finally, since we assumed that our concept is of a comparative nature, our empirical testing will involve finding out whether individuals can meaningfully order others on some scale so far as relative statuses are concerned. Measurability, in this case by means of scaling, is an important test of the usefulness of the concept in sociological research (Dumont and Wilson, 1967; Gould and Kolb, 1964; Weeks, 1972).

The fundamental issue in our discussion is whether full definitions of abstract concepts at the empirical level are attainable. Various attempts to achieve this have been equally unsuccessful. Let us take the example of the concept 'group cohesiveness'. An operationalist definition would rely on the *a priori* assumption of the researcher "as to the most crucial attractiveness factor for all group members, hence only one dimension of the innumerable dimensions of the total field of forces is utilized" (Dumont and Wilson, 1967). Reduction sentences which are conditional definitions, i.e. stipulating the conditions in which a particular characteristic can be attributed to an object, leave the definitional options free according to empirical circumstances, but are effectively specific instances of operational definitions. Here the assertion would be that a group can be attributed cohesive characteristics if it will respond by unified action Y (strike) in condition X (dispute between labour and management). Each specific reduction sentence will, therefore, provide only *partial specification* of the meaning of group cohesiveness. Another kind of correspondence is achieved where a whole theoretical sentence is translated into an observational sentence in order to test the former. In our example it will be true to say that "the more cohesive a group, the more its activities are valued by its members", if we find observationally that the sociometric measurements in the group showing a higher degree of positive interpersonal relations correlate with a higher frequency of activities selected within the group than outside it (Dumont and Wilson, 1967; Carnap, 1956).

In all these cases we are prevented from fulfilling the desideratum to provide as complete and integrated a definition of the concept as possible, which would move us in the direction of abstraction, by the equally pressing requirement that we relate the concept to specific observables, lest we allow ourselves to be plunged into an 'unlimited regress in definition'. In other words, however long the series of definitional steps, ultimately these must be linked to "something outside the realm of terms and symbols, namely to items of direct observation" (Feigl, 1949). An exception might be a mathematical definition such as an algebraic equation which defines some constructed class of patterns without this being linked to observables. The inherent logical primitives conventionally agreed-upon do not necessarily require empirical linkage. Yet individual manifestations of the constructed pattern will be expressed through the substitution of definite values for the variables in the equation pattern. What is more, the mathematical definition will achieve scientific functionality only where it is firmly linked to empirical phenomena. (Hayek, 1967; Quine, 1949; Hobson, 1968).

Neither this last statement nor the trend of our entire discussion should be taken as parts of a declaration of faith in strict operationalism. The stress on the ultimate need to link our concepts to empirical reality arises from a pragmatic approach developed within an inderteminist scientific framework. It must be clear, as Boltzmann has stated, that our concepts are mere images or symbols that have a *certain* relationship with reality but never *completely* correspond to it. There is no need to demur, therefore, from the idea of postulating unobservables. Certainly in the case of explanatory concepts, as opposed to purely descriptive concepts, the basis is often wholly or partly hypothetical, assuming "the *existence* of something which itself can be neither observed, nor readily be derived from observational data, nor yet easily reduced to empirical terms" (De Groot, 1968; Edel, 1979). But even in the realm of description we need not dispose of group or macroscopic concepts. The problem posed as to the undefinability of certain collective properties in observational terms is a moot point when we consider the irreducibility to observables of many a microscopic property. As Kaplan asks: "do we see genes when we look into a pair of blue eyes?". Or is a certain attitude of a single individual more given to empirical observation than, say, some decision taken by a group? In all these cases there are elements of hypostatization present. Yet logical coherence alone in holistic scientific constructs is not sufficient. They must satisfy the condition of applicability. To revert, therefore, to the earlier argument, we must shift the accent from questions as to how to arrive at the exact meaning of a concept, onto the issue of whether the concept is of empirical significance. The crucial question then is: in what way does the concept contribute to the explanation and prediction of observable events? For, in the last analysis, concepts are evaluated in terms of their efficiency; that is to say, one asks whether an explanation using a concept provides control over the empirical situation to which the explanation is applied. (Hobson, 1969; Brodbeck, 1968; Meehan, 1968; Dumont and Wilson, 1967; Edel, 1979; Kaplan, 1964).

Conclusions

Having considered conceptualization from various angles let us now conclude, with the aid of a few examples, on a note which reminds us again that the success of the social scientist will depend, to a large measure, on his ability to match his theoretical concepts with the observables in the reality he investigates. He will encounter a whole range of difficulties. Firstly, many indicators are in their origin not linked to any concepts; for instance, where they come into existence simply because some statistical data may have been collected as incidental to some other activity. Attempts to justify their existence and 'fitting' them to some concept so as to make them 'useful' are often not valid because of a lack of correspondence between the empirical indicators and the variables which represent the theoretical concept. Secondly, ill-defined or loosely defined concepts are hardly likely to provide the conceptual guidance necessary in selecting indicators. Thus, we still lack agreement as to how we define 'poverty'. Is it a level of material possessions and services, a condition of relative deprivation, or an attitude? Similarly, for instance, there are fundamental conceptual problems in deciding what is 'health'. Is it "the absence of clinical disease, the absence of pain, peak physical and mental efficiency, or a personal sense of well-being"? (De Neufville, 1975). Thirdly, where we use multiple indicators to test abstract propositions we come up against the problem of the uneven quality of the indicators, as mentioned above, and the questions regarding the relationships, if any, between these indicators or the possibility of subsuming fractional indicators under more inclusive ones. Thus, how highly are income and housing correlated or in what sense, if any, are we entitled to regard occupation as an inclusive indicator subsuming income, style of life, education, housing and other indicators, when dealing with the social class variable? Answering these questions is essential where attempts are made ro reduce the number of indicators or to combine them into an overall index. Fourthly, we must bear in mind that indicators are surrogates that stand in the place of our variables. We can never measure the latter directly. We do not measure social class as such but assess its importance by measuring its indicators (Bauer, 1966).

But if variables must have surrogates in order to facilitate their measurement, they are crucial in two senses: (a) Variables are the basic elements making up the analytical models and theoretical propositions. In this sense they enhance our quest for causal analysis, which is the central concern of social science. (b) Variables act as mediating concepts between the highly abstracts conceptualization necessitated by science and indicator operationalization needed to deal with the concrete situation. Conceptualization at the variable level enables us to observe the rules of correspondence between the theoretical and the empirical. Where we are dealing with practical problems, with policy formation and social intervention, this close linkage becomes a high desideratum. In the applied field there is no possibility whatever of relaxing this requirement. A lack of correspondence between the theoretical and the empirical is bound to result in faulty advice and would lead to wrong decisions. It is apt to conclude, therefore, that for both the pure and applied facets of social research it is crucial to understand the relationship between concepts, variables and indicators, which together constitute the bridge between knowledge and reality.

Bibliography:

Abell, Peter, 1971: *Model Building in Sociology*, Weidenfeld and Nicolson, London, pp. 28f., 38, 80f., 82f.

Allardt, E., 1972: *A Frame of Reference for Selecting Social Indicators*, Societas Scientorum Fennicas, Helsinki.

Andrews, Frank, M., and Withey, Stephen, B., 1976: *Social Indicators of Well-Being*, Plenum Press, New York, pp. 5-18.

Bauer, Raymond, A., ed. 1966: *Social Indicators*, The M.I.T. Press, Cambridge, Massachusetts, pp. 45, 99.

Blalock, Hubert, M., ed. 1961: *Causal Inferences in Nonexperimental Research*. The University of North Carolina Press, Chapel Hill, p. 48.

Blauner, Robert, 1964: *Alienation and Freedom*, University of Chicago Press, Chicago.

Brodbeck, May, 1968: "Methodological Individualisms: Definition and Reduction" in Brodbeck, M. ed., *Readings in the Philosophy of the Social Sciences*, MacMillan, London, pp. 280-303.

Campbell, Angus, Converse, Philip, E., and Rodgers, Willard, L., 1976: *The Quality of American Life*, Russell Sage Foundation, New York, pp. 13-17, Chapters 3, 14.

Carnap, Rudolf, 1956: "The Methodological Character of Theoretical Concepts" in Feigl. H., and Scriven, H., eds., *Minnesota Studies in the Philosophy of Science*, Vol. 1, University of Minnesota Press, Minneapolis, pp. 38-76.

Cicourel, Aaron, V., 1964: *Method and Measurement in Sociology*, Free Press, New York.

Costner, Herbert, L., 1969: "Theory, Deduction and Rules of Correspondence" in *American Journal of Sociology*, Vol. 75, 1969, pp. 245-263.

Curtis, Richard, F., and Jackson, Elton, F., 1962: "Multiple Indicators in Survey Research" in the *American Journal of Sociology*, Vol. 68, No. 2., pp. 195f., 204.

De Groot, Adriaan, D., 1968: *Methodology-Foundations of Inference and Research in the Behavioral Sciences*, Mouton, The Hague, Paris, pp. 37ff.

De Neufville, Judith, Innes, 1975: *Social Indicators and Public Policy*, Elsevier, Amsterdam, p. 121.

Dumont, R.G., and Wilson, W.J., 1967: "Aspects of concept formation, explication and theory construction in Sociology", in the *American Sociological Review*, Vol. 32, 1967, pp. 987, 985-995.

Edel, Abraham, 1979: *Analyzing Concepts in Social Science*, Vol. 1, Transaction Books, New Brunswick, N.J., pp. 19, 77.

Etzioni, Amitai, and Lehman, E.W., 1967: "Some dangers in 'valid' social measurement" in *The Annals of the American Academy of Political and Social Science*, September, 1967.

Feigl, Herbert, 1949: "Some Remarks on the Meaning of Scientific Explanation" In Feigl, H., and Sellars, W., eds., *Readings in Philosophical Analysis*, Appleton-Century-Crofts, New York, p. 500.

Forcese, Dennis, P., and Richer, Stephen, 1973: *Social Research Methods*, Prentice-Hall, Englewood Cliffs, N.J., p. 33.

Gould, Julius, and Kolb, William, L. eds. 1964: *A Dictionary of the Social Sciences*, Tavistock Publications for UNESCO, London, p. 692f.

Hayek, F.A., 1967: *Studies in Philosophy, Politics and Economics*, The University of Chicago Press, Chicago, pp. 23, 24.

Heisenberg, Werner, 1955: *The Physicist's Conception of Nature*, Hutchinson & Co., New York.

Hobson, E.W., 1968: *The Domain of Natural Science*, Dover Publications, New York, pp. 45, 49.

Kaplan, Abraham, 1964: *The Conduct of Enquiry*, Chandler, San Francisco, Calif., pp. 48, 53, 55, 71, 79.

Krausz, Ernest and Miller, Stephen, H., 1974: *Social Research Design*, Longman, London, p. 16.
Lazarsfeld, Paul, F., et.al., 1968: *The Uses of Sociology*, Weidenfeld and Nicolson, London, p. 167.
Lazarsfeld, Paul, F., 1973: *Main Trends in Sociology*, George Allen and Unwin, London, pp. 12, 14.
Lazarsfeld, Paul, F., and Reitz, Jeffrey, 1975: *An Introduction to Applied Sociology*, Elsevier, Amsterdam, pp. 84-86.
Meehan, Eugene, J., 1968: *Explanation in Social Science – A System Paradigm*. The Dorsey Press, Homewood, Ill., p. 85.
Merton, Robert, K., 1957: *Social Theory and Social Structure*, Free Press, New York.
Quine, W.V.O., 1949: "Truth by Convention", in Feigl, H., and Sellars, W., eds., Readings in Philosophical Analysis, Appleton-Century-Crofts, New York, p. 266.
Shonfield, Andrew and Shaw, Stella, 1972: *Social Indicators and Social Policy*, Heinemann for the Social Science Research Council, London, pp. 15, 16.
Warren, Richard, D., Fear, Frank, A., and Klonglan, Gerald, E., 1980: "Social Indicator Model Building: A Multiple Indicator Design", *Social Indicators Research*, Vol. 7, Jan., 1980, pp. 269f.
Weeks, David, R., 1972: *A Glossary of Sociological Concepts*, The Open University Press, Milton Keynes, pp. 26, 65f.
Zetterberg, Hans, 1965: *On Theory and Verification in Sociology*, Bedminster Press, Totawa, N.J.

Chapter 3

Causality and Social Phenomena

In treating the notion of causality our standpoint must be made clear from the outset. As social scientists we must strive for more than descriptive presentations in the form of frequency distributions or correlational data for variables. Our aim must always be directed toward causal analysis and explanation. We may invoke here an applied perspective, namely that our ability to remove some of the ills of society and to engage in social intervention through planning is contingent on revealing past causality and on predicting social events and their consequences.

Causal modelling is the main pivot of this striving for scientific explanation and it involves us in thinking in terms of 'antecedents', 'determinants' and 'independent variables' (Rosenthal and Rosnow 1975); or, as Blalock points out, we must accept the idea of 'forcing' or 'producing'. In other words we shall make statements of the type "a change in X produces a change in Y" and not merely of the type "a change in X is followed by or associated with a change in Y" (Blalock, 1964). It is true that in non-experimental research we can never demonstrate this causation empirically. But even in experimental research we may be unaware of a property that is the *true causal factor* in what appears to be an empirically tested cause and effect. Such limitations need not detract from the usefulness of theoretical causal explanation.

Causal explanation: basic notions

Causal explanation rests on logical inferences arrived at by testing hypotheses related to empirical generalizations and on 'hedging our bets' by means of stipulating 'initial conditions' and invoking the rule of *ceteris paribus*. We shall not concern ourselves here with an elaboration of the hypothetico-deductive method (Ryan, 1970; Hempel, 1965); nor with the full criticisms levelled at this method (Meehan, 1968; Hesse, 1976). A few remarks will be necessary, however, to signify the limitations that surround causal explanation. Thus, it is all very well to state that the requirement for successful explanation is that the premises should be fully corroborated. The fact of the matter is that we always face an insoluble *problem of induction* when basing ourselves on empirical generalizations. An empirical generalization presents a contradiction in terms, for if 'empirical' refers to description or observation and the term 'general' refers to a proposition that is valid without reference to any specific time and place, then no proposition can be both empirical and general. As E.J. Meehan says, the stumbling block is that "no amount of empirical testing could possibly establish a general proposition because it refers to the future as well as the past, to what has not been observed as well as to what has been observed" (Meehan, 1968).

What we are up against is the problem that universal statements can never be *fully* verified, however numerous the singular statements adduced for their verification. It is for this reason that Popper suggests for the demarcation of science the falsifiability or refutability of universal scientific statements. That is to say, scientific statements must be open to the possibility that scientific testing or everyday evidence may refute them, something that could not apply to metaphysical statements (Popper, 1961; Popper, 1976). Applying deductive logic in this sense would mean that a proposed theory or an existing scientific law will lose its veracity if contradicted by a single statement based on empirical evidence. The outcome is not always clear-cut however. Contradictory assertions may persist, especially in the social sciences, where the evidence might not clearly refute any of the contradictory theories. Alternatively, something that often happens in the natural sciences, the contradictory evidence may only require that some auxiliary hypothesis be modified or changed, and, therefore, the law will not in itself be falsified by what at first appears to be contradictory evidence (Lakatos, 1970). Hence Popper takes refuge in the uncertain world of provisionalism. That is to say, however logically clear-cut the deductive method might appear, it can only lead to statements, theories, or laws, whose truth always remains provisional and will never be fully proved.

Another apparent weakness of the deductive method is the distinction made in this paradigm between the *explanans*, the explaning element, and the *explanandum*, the event to be explained. In a sense this distinction appears to be artificial sinse the *explanans* cannot be conceived except through the *explanandum* (Cohen, 1962). Thus, we say that a road accident was caused by a mechanical failure making the vehicle career off course, these being the explaining elements, the *explanans*. But on reflection, careering off and mechanical failure are conceived as *causes* of the accident on account of the accident having taken place. Had the accident not taken place, they might have produced an infringement of the law or just a nuisance to the driver. Therefore, the same elements are now conceived as causes of law infringement or nuisance on account of these latter events, rather than causes of an accident. That is to say, we conceive the *explanans* through the *explanandum*, the event to be explained. However, this objection has little force. Although it *is* the accident itself which leads on to a causal explanation, when it does not take place it is because some initial condition is absent. Of course, whenever a phenomenon does not occur there is the convenient 'escape clause' of some unknown initial condition missing. This brings us to the issue of the *ceteris paribus* rule which, if wantonly applied, by leaving many relevant variables out of the picture, results in explanatory schemes with so many exceptions that they cease to be useful when applied to reality (Blalock, 1976).

The critical points made above bring to our attention the problems which arise when applying the deductive paradigm to the complexities of the real world. Put more sharply, the paradigm is constricting because it sets up a high standard of explanation at the level of logics which in reality is unattainable. On the other hand, the deductive paradigm has the merits that: 1. it focuses on a crucial principle of causal explanation, that is the observing of the regularities of the physical and human world; and 2. it insists on what may be called *causal necessity*, that is to say, that given certain conditions and some particular combinations and relations between sets of variables, the outcome must be X rather than Y. This is most important, since it means that we must be able to show why things turn out one way rather than another. Herein lie the central aims of all scientific inquiry: to understand, to anticipate, and to control (Meehan, 1968; Ryan, 1970).

Types of causal model

The principal aid for attaining these aims is the causal model. A model may be defined as "a realistic representation of something, like its subject in some ways and unlike it in others, often leaving its relation with its subject uncertain" (Harré and Secord, 1972). A scientific model reifies and simplifies the reality. As such it is isomorphic with the important features of reality. It is not a complete replica of reality. What is the function of such a scientific model? Since the aim of science is to explain by what and how the non-random patterns of life are produced, the causal model stands in for the causal mechanism supposedly at work. An hypothesized explanation for the unknown mechanism in reality is thus provided by the model, and is checked by scientific research.

It is important to distinguish between: 1) models of the inconic paramorph type which are scientifically creative, and 2) models which are homoeomorphs and have only heuristic value in science. An explanation of the difference between the two types must be given and some illustrations provided. Iconic models represent real or imagined things and are paramorphs when their source and subject matter differ. The virus is an example of such an iconic model. Its source was the bacterium and it differed from its subject matter which was the unknown mechanism that produced a certain disease. Harré and Secord see this kind of model as scientifically creative. On the other hand, a schematic diagram depicting, say, the nervous system, has as its source *and* its subject the same real nervous system, which means that it is a homoeomorph. It is valuable for understanding the structural relations between sub-systems and individual nerves, but it is not creative in the sense in which iconic paramorphs are (Harré and Secord, 1972).

Let us turn our attention now to causal models in sociological research. We can define a causal model in this specific case as *an analogue of some social process showing a causal interrelationship between the various elements involved.* We shall follow the distinction made by Harré and Secord by illustrating from sociology the two different types of model. Thus, the organicist-functionalist type of models, which have had a strong impact on sociologiacal research, are paramorphic iconic models. The organiscist model draws analogies between the biological organism and society or social groups, and uses concepts such as 'function', 'growth', 'system', and 'pathology'. The aim is to discover how society or a particular social group within it (both these entities often called 'social systems') works, and what are the functions of the component parts of such social systems. A case in point is Parsons' social system model, which can be equally well applied to micro studies, such as of peer groups, and to macro studies encompassing whole societies (Parsons, 1954). The only difference is that in the former the components of the system are micro variables, say personality systems, whilst in the latter they are macro variables, say institutional systems. The main criticism leveled at the Parsonian model is that it has transformed the *dynamic* biological model, postulating functional selection processes, into a *static* functionalist model when applied to the social system, or at best into a model allowing for process change.[1] This transformation has meant that the Parsonian model has ignored the important element of conflict as a cause of the dynamic changes taking place in social systems. As such, organicist-functionalist models have not been creative in explaining certain aspects of society, such as abrupt changes, violent revolutions, and the like.

On the other hand, one can glean the potential scientific creativity of other biologically oriented models. For example, epidemiology provides the iconic model whose source is the spread of disease for a subject such as rumour transmission. Sociological research would have to check, by using such a model, whether the process of rumour spreading resembles that of disease spreading (Brodbeck, 1968). Another example of an iconic paramorph type of model which has been creatively used in sociology is the model of 'the market situation', whose source is classical economics. This has been applied in research concerned with the sociology of religion. Peter Berger has analyzed the place of religion in the modern pluralist secularized society in term of a free competitive market situation. In such a situation different

[1] For a distinction between static, process, and change models see: N. Smelser (1968).

religious groups compete, in an ecumenical atmosphere, and regard the laity as consumers to be won over to their religious beliefs and denominations (Berger, 1969).

We shall give two major examples of homoeomorphic types of model: the Weberian ideal-type construct, and the mathematical model, with special reference to the path model. Weber's ideal-type construct is an abstraction of some one aspect of reality, such as bureaucracy or patrimonialism, providing in the generic senes a pure or perfect type of such a phenomenon. It is obvious that social phenomena either never attain such purity or can never be observed as such. But the construction of the pure or ideal type provides models for the analysis of reality and are in this sense imputed with heuristic value. Thus, the ideal-typical capitalistic productive system is abstracted out of certain features of modern large-scale industry, and can then be applied to study concrete cases of capitalist industry (Weber, 1949). The advantage of this kind of model over paramorphic models is that in seeking out the underlying elements and processes operating in society, society *sui generis* provides the heuristic device for its own dissection and study. Despite Weber's claims on behalf of the ideal-typical method, he made little use of such models in his substantive works, except for his study of feudalism in the later Middle Ages (Andreski, 1984). Nevertheless, it is interesting to note the claims made by Weber concerning ideal-types. There is no doubt that Weber saw his ideal-types as more than aiding critical scientific description. Weber stresses that the ideal-type does not merely highlight the 'significant components' by comparing the empirical arrangements and sequences of events with the logically constructed system. It also provides "an analytical conceptual framework for our understanding of these empirical arrangements and events in *space, time, and causal relationship*". In other words, he sees the ideal-type as "the means for explicitely and validly imputing an historical event to its real *causes*" (our italics) (Weber, 1949).

The question is how a proposition to be tested or a theory formulated, as to the causes of a social phenomenon, are arrived at by means of the ideal-type? An example will be in place here. We can take Robert Redfield's characterization of the folk society as an ideal-type, which has the following features: "small, isolated, non-literate, and homogeneous, with a strong sense of group solidarity. The ways of living are conventionalized into that coherent system which we call 'a culture'. Behaviour is traditional, spontaneous, uncritical and personal; there is no legislation or habit of experiment and reflection for intellectual ends. Kinship, its relationships and institutions, are the type categories of experience and the familial group is the unit of action. The sacred prevails over the secular; the economy is one of status rather than of market" (Redfield, 1947). Redfield himself and other investigators have compared this ideal-type with actual societies and have found different degrees of approximation with the properties of the ideal-type. Such studies have also suggested how these properties are related causally, and have led to general propositions such as that in a folk society there is "no disposition to reflect upon traditional acts and consider them objectively and critically" (Redfield, 1947). It appears that the heuristic value of such models lies in the *suggestions* provided by a comparison of the ideal-type with the real world situation. These suggestions can lead to theoretical propositions. But this is a different kind of modelling from that where paramorphs are used. In the case of the latter the theoretical proposition is directly applied from the model whose source is another area of reality. To state this more clearly: in the case of the ideal-type the theoretical proposition is a *consequence* of the application of such a model to the real world; in the case of the iconic paramorph the model already contains a theoretical proposition and its application consists in testing the theoretical proposition in the real world situation, which is the subject of the research.

Mathematical models provide the second major example of homoeomorphic models. In recent decades they have been more and more resorted to in sociological research. Mathematical models are quantified presentations of the hypotheses relating to the data of an investigation and the evaluation of the relationships proposed by the hypotheses. This is true whether the model is a simple two-by-two table, a complex multivariate analysis, or a diagram, each showing in quantified form the strength of the relationships between the variables. What must be clear is that the building of any such model requires certain preliminary steps: 1) the classification of the variables into independent, intervening, and dependent ones; 2) the drawing of tentative arrow diagrams to indicate how the variables may be related (Galtung, 1967). But such preliminary steps are themselves based on some theoretical assumptions regarding the causal relationships extant in the relevant reality. These assumptions or hypotheses must exist *before* the mathematical model is built and used. In other words, the researcher has in his mind an image of how the processes run, of what causes what, and so on. We see, therefore, that no model or model building process can be atheoretical. This applies to ideal-types as well as to mathematical models. But the homoeomorphic model helps us arrive at a full theoretical proposition only by its application to reality. In contrast, the paramorphic model provides a full-blown theory from one subject area of the real world to be used in another subject area.[1]

We shall give below the example of a mathematical path model and advance some of the criticism levelled at that model. As for the moment it must be stressed that mathematical models can be productive providing: (a) the

[1] We discuss below the relationship between theories and models.

assumption of measurability/quantifiability holds for the topic investigated; and (b) the properties of the model are applicable to the reality investigated. Thus, not always is quantification feasible and relevant. The sort of data that can be elicited is not always suitable for exact measurement. A case in point could be W.F. Whyte's study of primary relations in street corner gangs (Whyte, 1943). Again, some mathematical models have properties such as linearity or asymmetry, which make them unsuitable for studying certain societal aspects of life. Thus, aspects of social interaction or conflict cannot be handled by linear models. As Peter Abell says, sociology has relied heavily on models based on the assumption of linearity, but since non-linear convariation patterns are probably very common, a change towards models which have such properties can be expected (Abell, 1971).

Mathematical modelling on the whole has achieved greater sophistication over the last two decades. Thus, in the multivariate path model the existence of 'residual' and 'intervening' variables is recognized, and further refinement is added in stochastic models where the probabilistic nature of the relationship between variables is stressed and a multiplicity of possible outcomes is presented. It may be noted here that the inherent advantage of a predictive stochastic model is that it can deal with the element of uncertainty that characterizes much of human behaviour, due to the possibility of choice available to individuals. What probability theory does is to accommodate this choice factor by transforming individual choice into a chance factor when overall variability in the group is described. Such stochastic models may be used not only to deal with past descriptions of transitions between given positions over a period of time, such as in the case of intra-generational social mobility presenting transition rates between occupations. They can also show expected inter-generational mobility, thus providing a forecast of the future occupational structure. The goal is obviously to take account, in such models, of possible situational or condition changes (Coleman, 1973; Bartholomew, 1973). Such predictive models can provide the likely results of different policies and the likely effects of different conditions.

The problem of causal modelling

We have, so far, focused on the nature of models and the types of model used in sociology, having initially stressed the indispensable role that *causal* modelling plays in social research. A word will be necessary here about the principle of causality itself and the problems associated with it. The question is often posed as to how we actually determine the direction of causality, where correlational evidence between two variables is revealed. In some cases commonsense suggests the direction of causality. For instance, in the case of constant association, in different countries and over many years, between low marriage rates and periods of economic depression, it is pretty clear that the latter causes the drop in marriages rather than *vice versa* (MacIver, 1964). It is often the case that temporal sequences will indicate for us the direction of causality. This means that a particular point in time is taken as providing the 'initial conditions' from which our causal determinants emanate to *produce* some result. Much depends, therefore, on the cut-off points, so that if we accept, as an example, that the Protestant Ethic gave rise to the spirit of capitalism, this means that our cut-off point is sometime during or after the gestation period of Protestantism. This is arbitrary, since we could have chosen a period before Protestantism appeared and, hence, we could have lengthened our causal chain and pinpointed some other earlier determining factor, say some practices of the Roman Church or some changed economic needs in medieval Europe. Similarly, where a close interdependence occurs between a large number of variables causal ordering will be extremely difficult.

In addition, there are also the phenomena of feedback, contamination and reciprocity of which the social scientist, undertaking causal explanation, must be constantly aware. The reciprocal or circular causality may be illustrated by the following example: an influx of immigrants brings about public pressure for the imposition of restrictive entry legislation; prospective restrictive legislation results in heightened immigration due to 'a rush to enter before the doors close'; in turn, this larger influx further strengthens the pressures for tougher legislation. Circular causation is, therefore, taking place. As far as contamination is concerned Duncan raises a serious problem. Thus, taking a model of the process of achievement utilized to interpret data from a cross-sectional study in which indicators of the motivational variable are secured, as of the time of the study, we cannot assume that it is possible to collect uncontaminated retrospective data regarding this variable, as is the case for, say, socio-economic information. Since the motivational variable is subject to contamination by the dependent variable in the model, no causal inferences are warranted unless we can draw on additional information and involve further assumptions. To do so, however, would introduce further complications into the model (Duncan, 1969).

Indeed, attempts have been made in mathematical and diagrammatic models to handle complex relationships by means of recursive and feedback systems, by taking account of indirect effects, by the decomposition of variables into their components, by stochastic analysis, and so on. Nevertheless, serious reservations remain concerning such models. Take the path model which basically provides "an operational explication of the notion of deductive theory". Its focus is on linear causality and the causal interpretation of statistical relationships. It involves "dependence analysis" and it assumes "a particular causal ordering of the variables involved". The "linear, additive, asymmetric relationships among a set of variables . . . are conceived as being measurable". The diagrams used have not only a mnemonic function; they also carry notations of the correlation coefficients, so as to indicate the strength of relationships between variables (for elaborated presentations and examples of diagrams and equations see: Duncan, 1971; Turner, 1964; Duncan, 1975; Blalock, 1971).

Now, such models are not only unsuitable for handling certain types of social relationships because of the property of linearity, as mentioned above, but their interpretation of data rests on assumptions about the ordering of the variables and about unmeasured variables, which are treated as 'residuals', thus showing up the path model for what it really is, i.e., a partial structure. This is so, since the structure presented by such models includes variables that account for only a fraction of the variance in the dependent variable that we seek to explain. We must assume, therefore, that the 'residuals', i.e., the fraction of variance unaccounted for, and the source of which is not known, emanates from independent variables not important enough to change the entire structure of the model. This would mean that although we have, in fact, only a partial structure, it is sound enough to retain explanatory power; an argument most difficult to accept. Duncan defends the path model on the grounds that "it makes the assumption explicit and tends to force the discussion to be at least internally consistent", and that "with the causal scheme made explicit, it . . . enables criticism to be sharply focused and hence potentially relevant not only to the interpretation at hand but also, perchance, to the conduct of future inquiry" (Duncan, 1971). But this defence does not take care of more fundamental objections.

Ray Pawson provides some of the serious caveats. Thus, the path model rests on a logic which forces us to separate the individual from institutional constrains, the latter not being definable in such a causal model. He exemplifies through Blau and Duncan's path model concerning the American occupational structure (Blau & Duncan, 1967). To take just one instance, the explanatory logic of this model tells us that the relationship between the father's occupational status and the son's occupational status might be one of a number of relationships that determine occupational status. But as Pawson says, "What this particular logic ignores is that this association also depends on the occupational structures of the society in which the placement occurs". There are other problems too, such as the inability of the model to accommodate different meaning structures. For instance, is occupational status similarly valued, interpreted, or regarded in the various sub-cultures of a society? Are causal modellers entitled to assume that they "deal with events which demonstrate a uniform cultural significance", and apply this model as if a certain universality pervades populations? Pawson argues that in effect path analysis is no more than a snapshot providing us with "a set of momentary regularities literally plucked out of a much larger explanatory structure" (Pawson, 1978).

Figure I

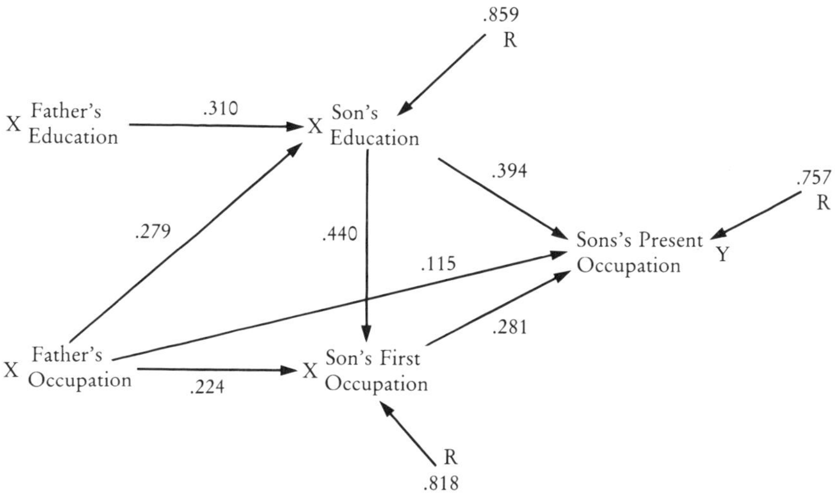

X = known causal variables
R = residuals, i.e., unknown variables
Y = dependent variable

Adapted from Blau and Duncan's model of the AMERICAN OCCUPATIONAL STRUCTURE

The question is whether the above mentioned limitations, often subsumed under the umbrella argument of 'the complexity of human actions', and other difficulties encountered in the social science field, such as the contextual uniqueness of social events, will ultimately defeat our endeavours at causal analysis (Dixon, 1973). It appears reasonable to suggest that these prevent us from taking a thoroughgoing deterministic line. For it is clear that in real-life situations we shall never be able to take account of *all* the possible connected variables; neither shall we be able to assume the complete isolation or non-contamination of any social events from other occurences. Causal social chains, looked at metaphysically, are as long as human history and their extension into the future is endless (Russell, 1954; Hanson, 1965); and it is an equally daunting prospect for the social scientist to contemplate the myriad of interrelationships between social events. At the causal analytical level this complexity of interrelationships shows itself in terms of 'multiple causation' and 'compound causality', meaning that events may have several causes working either in parallel and independent manner or in interlinked combinations, that is to say through mutual dependency and interaction (Dixon, 1973).

But neither the taking of an indeterministic position nor the difficulty, as we shall see below, of producing 'law-like' sociological generalizations, need deter us from our job as social scientists to describe the social world around us and to propose causal explanations of the events described. It has to be admitted, though, that causal explanation is often circumscribed contextually. Another limitation, not unlike that found in the natural sciences, is that the explanatory scheme includes only those variables deemed immediately relevant. Finally, the social scientist can never expect to achieve perfect knowledge of the world he studies. Hence his analysis is always subject to error. Notwithstanding all these limitations, he cannot give up his quest for causal explanation. This is so, whether his aim is purely to enhance social scientific knowledge, or his purpose is intervention and experimentation in order to solve real-life problems.

Theories and laws in social research

Next we must consider, albeit briefly, the relationship between causal models, theories, and paradigms. Earlier on we dealt with the nature and types of causal model, but have not clarified how an actual model is arrived at. To put it simply, on what grounds do we build a model that postulates a causal link between X and Y rather than one that postulates a link between Y and Z? Lave and March suggest that model building is a form of art excercised through speculative thinking about the observed world around us. In their broad view 'model' is a "generic term for any systematic set of conjectures about real world observations" (Lave and March, 1975). Others, on the other hand, stress the logical relationship between scientific theory and the causal model. According to this view models are well grounded in substantive theories which generate them. The relationship, however, is a mutual one, for the model checks on the substantive theory and reformulates it (Land, 1969). Furthermore, some models can be the basis for and can develop into theories by means of empirical corroboration. Thus, Weber's ideal-type bureaucracy, which is a model, can be developed into a theory of the bureaucratic structure and process by using the model to actually test the real world. The distinction, therefore, is not very sharp, for both the model and the theory assist us to sort out causal relationships. But whilst the model is a creative or heuristic device, the theory makes a claim based on empirical evidence. In other words, the model belongs to the context of discovery, theory to that of validation (Ryan, 1970; Rudner, 1966).

It would appear from our immediately preceding discussion that the guidelines which we obtain from existing theories in order to build our models are reduced to the level of hypotheses once incorporated into the model. The latter, then, provides the basis upon which we test, modify, and crystallize new theoretical propositions. For instance, based on a theory of adaptation, asserting that individuals or groups modify their behaviour in response to experience, we might propose the use of a 'game model' in our research project (Lave and March, 1975). This in its turn will help us put forward hypotheses, becoming theoretical statements once linked to empirical evidence.

There seems little doubt in the minds of most sociologists that theory formation, and therefore causal explanation, is a primary task in sociology. Many reservations and doubts have been expressed, though, concerning the success that sociology has had in this field. Most writers point to the paucity of 'law-like' generalizations or general theories. There are, of course, certain *constants* or universal features of social life which can be regarded as basic 'law-like' generalizations, such as the interrelationship and interdependence of individuals in groups. Sociology has also produced overall models of society and conceptual frameworks, which Merton calls 'grand theories', such as structural-functionalism or symbolic-interactionism. The self-criticism of sociologists concerns, however, the inability of the discipline to produce general theories about *specific aspects* of society. There does not appear to exist, for instance, a general theory of deviant behaviour. Instead, we seem to encounter broad statements of trends or statements of statistical correlation backed by plausible interpretations (Dixon, 1973). At best, it is argued, sociology has produced, in the field of deviant behaviour and in most other fields, "single determinant theories" (Emmet and MacIntyre, 1970). The reason for this, according to Georges Gurvitch, is the discontinuous nature of social events, a view difficult to accept; although even Gurvitch maintains that "sociological causality which is singular, never goes to the extreme of the singularity" espoused in historical causality (Balandier, 1975).

Against Gurvitch's view one could stress the regularities in the social order, the everyday patterns of activity, the links in the social structure, and the continuous nature of society. Sociology could in fact be regarded as nomothetic. As Erikson notes, however, human actions and social events have partly generic and partly specific properties: it *is* possible to categorize them, yet in some ways they remain unique (Erikson, 1973). Most sociologists lean towards categorization, seeking out the social patterns and seeing the links and continuities. They do not ignore the contextual aspect, but the stress is on the *social* context and on the comparative analysis of the various social contexts. But there are those few who stress the uniqueness of social actions and who strive to understand the meanings of single social situations (Mennell, 1974).

It is often maintained that the corpus of sociological knowledge shows causal explanation not to have reached a level necessary to establish 'law-like' generalizations. Can it be that the basic constants or universal features of societal life, that we mentioned above, are more than counteracted by the complexity of human actions, the interdependence, interaction and mutual influence of groups and events, and also by the specific contextual aspects and the discontinuities? Does this mean that we are prevented from uncovering true causality and from establishing sociological laws? Not all would agree with such a view. For instance, Louis Guttman claims that universal laws *can* be established in the social sciences. He claims, for instance, that there is a law concerning certain attitude patterns discovered in usual populations, this being an area which is often connected with sociological research.

This search for laws, which is the *raison d'etre* of most scientific work, normally upholds the principle of causality. The yearning for universal laws and for causal understanding is expressed by Karl Popper thus: "We are not to abandon the search for universal laws and for a coherent theoretical system, nor even give up our attempts to explain any kind of event we can describe" (Popper, 1961). This exhortation contains two distinct goals which need not be connected but which often do appear together: cause and effect on the one hand, and regularities on the other. Let it be clear, therefore, that in principle X could cause Y, but not occur more than once, hence regularity is not involved. But regularities are logically not independent of the causal principle, since one should be able to explain why there are no variations from the regular pattern, i.e., what causes the regularity. Yet it is true that regularity may be observed without knowing its causes. This in turn leads to the question of how one can know whether certain occurrences are absolutely regular: in other words, is observation alone sufficient to establish the existence of regularity? Furthermore, how is our explanatory power affected by the factor of chance? Scientific work does not offer clear-cut answers to such questions, having instead leaned heavily on the problematic principles of 'inductive probability' and the 'primacy of repetitions'. That is to say, since the absolute truth of a law cannot be established, its probable truth will receive inductive support from repeated evidence, which means accepting the doctrine that "repeated evidences furnish a kind of *justification* for the acceptance of a universal law" (Popper, 1961; Reichenbach, 1951; Strawson, 1970; Lessnoff, 1974).

It is precisely on this principle of repetition that Guttman bases his assertion that in at least two areas of human behaviour universal laws have been discovered: they concern intelligence and attitudes. Thus, repeated studies, over many years, have shown that "correlations between mental tests almost always turn out to be positive", whilst the results of numerous researches show "that attitude items toward a given object are not negatively correlated for usual populations' (Guttman, 1978; Gratch, 1973; Shye, 1978). Guttman and his associates provide us with highly interesting findings, which may prove constructive steps in the search for laws in the social sciences. Yet the limitations of these 'laws' should be quite obvious. Firm as they seem to be, appearing to have the character of underlying 'natural laws', these are not *causal* laws. Guttman's 'laws' are really behaviourist descriptions of certain regularities, these regularities appearing in the form of scalable attitudes or ranges of intelligence. We do not know from such 'laws' what are the causes of such regularities. Guttman's system sets up the necessary conditions for the formulation of universal descriptive laws, i.e. regular occurrences and typologies, but it does not provide *sufficient conditions* for arriving at causal laws. No teleological argument is implied here, since it is not a question of seeking final or ultimate causes but rather the immediate, first order, explanation of certain recurrent phenomena. Perhaps it is totally irrelevant to raise the question of causality from the social scientific point of view, since these may be phenomena reflecting innate human characteristics. But whatever the sources of the regularities discovered, the validity of the proposed laws seems to depend on constant adjustments between the definitional system for observation and the technical manipulation of data. Thus, much depends on the definition of the population i.e., what is a 'natural' or 'usual' population, using their terminology. A case in point is the exclusion of a population with a characteristic that has a close affinity with the specific contents of one of the items. For instance, the law of attitudes was violated when negative correlations were found between the items "valuing being religious" and a series of items appraising the importance of other activities and experiences, in a sample of *religious* high school students. It is suggested, therefore, that the latter are not includable in a usual population, given the attitude universe in question. It thus follows that not only the definition of the population is crucial, but also the decision regarding the attitude items to be included in the test.

As Samuel Shye clearly states, there is need for "further research [to] help clarify what constitutes a natural population relative to an attitude universe". How is this to be achieved? Apparently by "continued testing through experimental replication, so that any one or both of their parts, the definitional and the technical, may be refined, modified or altogether replaced". This approach reminds us of the methodological rules that made up the 'research programme', often employed to protect 'hard core' laws, if we look for instance at the history of the physical sciences. These rules, as Imre Lakatos tells us, consist of the *negative heuristic* which indicates the paths of research to avoid, and the *positive heuristic* which indicates the paths of research to pursue. This means that the *modus tollens*, the reasoning that removes the validity of a proposition, is directed away from the stated 'hard core' law and is shone onto ancillary hypotheses. This is precisely what Shye tells us to do when he suggests that where a law does not hold in a certain context, this will lead to "constructive sharpening of the experimental conditions and of the definitional framework for example by clarifying what constitutes a 'non-artificial' or a natural population in the context of intelligence testing or possibly by further restricting the definition of intelligence items" (Shye, 1978). Or in Lakatos' words: "[the] 'core' is 'irrefutable' by the methodological decision of its protagonists: anomalies must lead to changes only in the 'protective' belt of auxiliary, 'observational hypothesis and initial conditions' " (Lakatos, 1970). It may well be, Shye argues, that in this manner we arrive at only "the crudest law", but this is still preferable to no law at all.

The statisticalist view of causality

Despite the limitations, pointed to above, with regard to the discovery of law-like generalizations through statistical correlations, we shall now suggest that the statisticalist approach is, nonetheless, the most significant and viable attempt to uncover regularities and causality when dealing with social phenomena.

One of the reasons for the difficultues encountered by social scientists dealing with the phenomena of the social world stems from their inability to escape the determinist notion of 'causality'. We join those who argue that the notion of causality may be interpreted in different ways and that causality understood in terms of a universal determinism, modelled on the idealized physical sciences, no longer serves as a stipulative definition for all scientific research in the sense that Nagel suggests (1961). That is to say, it cannot provide the ground-rule for all scientific inquiry. What has happened is that the very concept of 'causality' had undergone a change and thus has acquired a new meaning. This new direction emerged from physical science itself, where certain empirical facts in quantum physics forced the scientist to present his causal interpretations by means of statistical probability relations. This led to an indeterminist model of causality, which opened up the possibility of using the concept of causality in its new interpretation also for the explanation of social phenomena.

A very brief review of the various definitions of causality, highlighting the difference between the determinist and indeterminist interpretations, will suffice here. One way of defining causality deterministically is to say that the fulfilment of certain *necessary and sufficient conditions* will always be followed by the occurrence of an event. It is not clear in this kind of definition how we distinguish the causal factor or factors from the rest of the necessary conditions, which together constitute the sufficient conditions. Be that as it may, what this kind of definition stresses is the apparent constant conjunction between cause/condition and effect, that is the unvarying succession involved (Hospers, 1956). Another, more rigorous, formula is: "If C happens, then (and only then) E is always produced by it". The latter definition does not assume that constant conjunction and constant succession show causation, since both could be accidental. In this formulation "the gist of causation is uniform production". Mario Bunge argues that even processes that are synchronous or functionally related do not show a causal bond. The latter exists only when A *produces* B (Bunge, 1963). Yet another way of defining causality in a strict determinist sense is to insist that certain causes *necessitate* an event absolutely, a case in point being an eclipse. It can be argued, though, that cases of necessitation of this sort do not force us to accept the determinist interpretation of causality universally. That is to say, other examples may be brought to show that the term 'cause' can be used in a different sense, so that events, outcomes, or phenomena are not seen to be necessitated in the absolute sense. Namely, the causal principle is applied probabilistically, which means that variability exists in the conjunction between cause/condition and effect, and thus *causation does not entail uniform production*. The indeterminist-probabilistic causal statement will read therefore: 'State X is followed by state Y with probability p, and by state Z with probability 1-p" (Rescher and Urquhart, 1971).

There are, however, some major objections against the possibility that such a probabilistic statement can actually reveal causal relations. The chief objection is based on the claim that there is a lack of examples one can draw from the social sciences which exhibit the kind of causal structure in which one can screen-off all spurious relationships and one can demonstrate causal priority. In other words, when we adduce correlated data, these will be open to different interpretations, since without experimental demonstration we can never be sure that A is the cause of B. All we can say with confidence is that A is in some way related to B, but statistical reasoning in itself will not provide proof of *causal* relations. This lack of confidence then stems from the fact that (a) we cannot actually *see* the causal 'mechanism' working i.e., we cannot show the existence of a built-in causal structure, and (b) that even where demonstration of the causal connection could be provided, we are still not sure that we have a complete account of the causal structure. In the physical sciences experimentation enables us to *see* causal connections, but even here many areas such as micro-physics and medicine are replete with incomplete accounts of causal structure.

Epistemologically, the question boils down to whether one can use the concept of causality in cases where the causal structure cannot be fully demonstrated. That is to say, if events show stochastic patterns and the only explanations we can arrive at are based on statistical inferences, does this prevent us from proposing that events are causally linked, even if only in the weaker statistical sense; namely, that A in *many* instances if not in *all* instances will be followed by B? Our argument is that the latter is admissible and that there is no warrant for reserving the notion of causality to the deterministic idea of events being absolutely necessitated.

We wish to illustrate the idea of causality arrived at through statistical inference. The work of John H. Goldthorpe and colleagues on social mobility in Britain is quite typical in this respect. Let us consider their major conclusions concerning inter-generational class mobility, which were based on the statistical results of their survey data: 1. the increasing rates of upward mobility over several decades were found to be associated primarily with changes that occurred in the structure of the social division of labour due to a period of sustained economic growth, making it possible for lower class members to move up mainly into a growing 'service class'; 2. obversely, the increasing rates of mobility were not seen to have resulted from changes in the degree of fluidity of British society. The factor thus most clearly seen as increasing mobility rates is the economic one, not the social, and more specifically as far as the latter is concerned, egalitarian efforts are not seen as a major force influencing mobility chances, but appear to have some significance as a countervailing force to class structure. Goldthrope also uses the data for predictive purposes, saying that in a bleak economic future a decline in mobility chances would have a particularly high probability. He says: "If ... a period of significantly reduced rates of economic growth were to occur, the same scope could scarcely exist as previously for the expansion of the higher levels of the occupational structure, nor thus for a comparable increase in opportunities for social ascent ... The most obvious outcome would be a levelling off, and quite possibly a decline, in the amount of upward mobility; and in particular, the very low relative chances of mobility from working-class origins to the higher levels of the class structure that we have demonstrated would become increasingly manifest in similarly low *de facto* rates. If one anticipates a rather bleak economic future then a *decline* in openness would seem to have a particularly high *probability*" (our italics) (Goldthorpe, 1980). It is clear that Goldthorpe uses his statistical data to draw causal inferences probabilistically, which we can sum up as: if E_1 then most likely M_1 (E_1 = economic growth; M_1 = high mobility rates), and if E_2 then most likely M_2 (E_2 = economic decline; M_2 = low mobility rates).

The Goldhorpe study, we have adduced, has obvious weaknesses as an example of a causal theory. 1. Although its findings are presented as causal, we cannot invoke 'priority' so as to separate cause and effect; that is, one could argue equally well that lack of social mobility is the cause of economic depression rather than *vice versa*. There is no need to repeat here the arguments around the Humean problem couched in terms of our inability to observe the *power* of one element to cause another, and whether this can be solved by demonstrating that changing the supposed dependent variable does not change the value of the causal variable.[1].

Clearly, even if this were sufficient, the context in which studies like that of Goldthorpe take place do not allow such experimentation, and further doubts remain because of our inability to deal adequately with the problem of spuriousness. Yet such probabilistic inferences are not thereby invalidated. For the interpretation of the data, with the aid of intuition and assumptions based on other existing information, suggests a causal relationship that one can accept as plausible. Goldthorpe's findings regarding the expansion of service class occupations is not linked directly to the phenomenon of economic growth to which he refers. The statement that the relevant period was one of sustained economic growth is based on other existing knowledge, the assumption being that this was associated with an expansion in service class occupations. Thus, the major inference based on the survey data, namely that this expansion accounted for a great deal of the upward mobility found, is aided by an auxiliary inference. But this does not diminish the plausibility of Goldhorpe's causal proposition.

2. Although we use the Goldthorpe example simply to show how we arrive at causal propositions by means of statistical data, another point may be raised about his data. This is that the findings provide evidence that can support only a 'local' theory. Whilst this is true, such a 'local' theory could, with many other examples, support a more generalized theory, say a theory of class conflict on Marxian materialist-determinist lines. But there are some social scientists who argue that since only "*partial* regularities, trends or *local* laws can be empirically observed", general laws are unrealizable, and that statements concerning the future can be only conjectural (Boudon, 1983). Nevertheless, it cannot be denied that regularities and covariations observed statistically do exist, and even if they are not universal, they do appear widely in certain conditions, e.g., post-war baby booms, declining church attendance in cities, increasing levels of prejudice with a larger influx of immigrants, and so on. Underlying such regularities there are, surely, generalizable causes.

Another objection to the idea of applying the notion of causality to the social sciences arises from the substantively different nature of this field when compared with that of the physical sciences. Stephen Turner formulates this argument clearly when he says that since social sequences are routed through the cognitive processes of the individual's human action, simple spatio-temporality breaks down, leading to a situation of multiple causes which are impossible to disintangle. Thus, if we attempt to analyze, say, the single delinquent act of a juvenile then, indeed, we must "consider its relation to the myriad of things which might be claimed to be its 'causes', from violence

[1] For examples of complex causal structures based on statistical data see: Stinchcombe, 1968.

in television, the decline of the West, 'bad influences', the Y chromosome, the neighborhood, yesterday's demeaning encounter with an authority figure, to the degree of opportunity. Not only are there a great many possible causal relations, but there is no way to use such requirements as spatio-temporal continuity to decide between them by automatically excluding most of them as 'remote' causal relations". The core of the problem, according to Turner, is "the problem of human action itself" (Turner, 1982).

One can counter this point by arguing that the very strength of the statisticalist approach resides in the possibility it offers of avoiding reference to individual processes or particular events routed through the mental realm. For it seeks to uncover the frequencies of certain attributes in particular reference classes which are deemed to be relevant to the problem or topic under scrutiny. Thus, we feel entitled to assume that the sociologist is not interested in the social mobility chances of any *one* individual, but rather in the mobility trends that obtain in certain populations in particular periods. Similarly, the sociologist is concerned with the trends in delinquency rather than with the causes of any one act of delinquency. However, it is true that even when we consider the substantive field from this aggregate point of view, we discover "elaborate patterns of highly correlated events". The various techniques for sorting out these complex patterns, such as randomization, quasi-experimental control, the detection and screening-off of spuriousness, do not ensure that we can account for all possible confounding influences (Salmon, 1967; Salmon, 1971; Papineau, 1978). These problems arise, however, not only where human cognition is involved. In many areas of the physical sciences, such as in meteorology or geology, we find complex systems and situations that have "a high degree of uniqueness" and lead to uncertain prediction (Blalock, 1984). In such areas universal causal explanations are difficult to attain. And it is true that in both the physical and social sciences, in the final analysis we must appeal to 'tutored intuition', even where experiments are used. It is intuition that enables us to present the more plausible connections and to suggest a cause and effect relationship between variables. This implies a degree of uncertainty, yet intuition aids rather than destroys the search for causal relations.

It would seem to us that the objections raised against the aggregate-holistic approach in sociology stem from several different standpoints. One is the view that human action is free, emergent and controlling, and consequently not given to causal categorization. Carol C. Gould expresses this most clearly: "The concept of doing something or of an action entails choice or that the agent initiates the action. Furthermore, the concept of choice entails that an agent could do otherwise. By contrast, causal explanation in any of the three senses ... productive, nomological or holistic-structural ... requires that an action be understood as the effect of antecedent causes, such that the agent does not initiate the action and could not do otherwise, given the initial conditions ..... Such causal accounts thus fail to do justice to the distinctive character of human action as free" (Gould, 1983).

Another standpoint is that the finding of general covering laws about human action *is* possible, but only in terms of the action of the individual. George Homans is one of the chief protagonists of this view. He does not deny the existence of social structures, but sees the latter as outcomes of individual action based on behavioural principles. These behavioural principles rest on the idea that rewarded action will be repeated, the obverse of which is the avoidance of or escape from punishment. Herein lies, therefore, the covering law of all social behaviour, and the social patterns and structures observable are mere outcomes of such behaviour; and no general explanatory causal law can be based on propositions concerning the larger social patterns and structures (Homans, 1982).

There is a third standpoint best represented by Raymond Boudon. It seems that Boudon wishes to espouse the importance of the individual's freedom of action, but does not wish to fall into simple behaviourist psychologism. For him, therefore, "individual action and the products of individual action constitute the only and ultimate reality with which the sociologist has to deal with"; but these individual actions take place in situations of interaction which are variable and immensely changeable. It is for this reason that general laws, trends and regularities have "so many exceptions, that one wonders in which sense they are still rules". Hence, Boudon goes for *ad hoc* explanations and the stress on singular contexts, with the aid of the general conceptual model of the "bounded situation of interaction", which puts constraints on individual action but allows a measure of freedom of choice (Boudon, 1983).

The above anti-collectivist approaches differ in certain respects but they also have much in common. They do not deny the existence and even importance of macro phenomena, namely: cultural contexts, inter-group relations, institutional arrangements and the like, but their claim is that the important thing is to understand what goes on inside the macro pattern, that is to say the micro events which maintain them and shape them, e.g., the model of reciprocity takes "as its fundamental entities social individuals who constitute the forms and structures of social life by their activity" (Gould, 1983). And from the point of view of our discussion here, the crucial point they press is that no generalized causal statements can be proposed about those macro patterns. Why is this so? Carol C. Gould, for instance, claims that the category of causality is inapplicable to human action since the latter incorporates in it at one and the same time both cause and effect, and since social life itself is constituted by such action, it cannot be analyzed in terms of the causality principle. Raymond Boudon's conclusions are similar when he pronounces laws, trends and regularities in the social sciences as false and pseudo-empirical since they are so often contradicted by reality itself. Therefore, one can rely only on the micro-system of action. In Boudon's words: "A relationship between A and B can be nothing else but the product of individual actions, so that A will produce B or its opposite depending on the complex set of characteristics of the system of action" (Boudon, 1983).

Such individualist premises lead to what we believe to be the key point about this approach. We refer to the insistence that the human agent is sufficiently autonomous, innovative and recalcitrant so as not to fit into invariable categories necessary to provide criteria of universality at the macro level. This means that causal laws cannot be proposed *sui generis* at the level of aggregate human phenomena. Thus, for Homans generalizations at the aggregate level always stem from causal laws at the micro level, whilst for Boudon the aggregate level does not endure true generalizations but only analysis in the singular context. The upshot is that, on this view, the only stable aspect of the shifting sand of human action, the only universal feature, is the framework and technique used in the micro-exchange situation, namely the reciprocity type of relationship, the action-reward system, the strategic interaction or communication, and so on. If this is the message of the individualists, and we believe it is, then their contribution to the discipline of sociology is trivial (see an example below). We also think that underlying this position is the mistaken notion that the universality thesis must be the basis of the principle of causality, i.e., the assumption that "there is always an invariable law that will cover a causal sequence". Deborah Rosen in a strongly argued paper criticizes this deterministic assumption concerning causality and denies that "causal relatedness is in its nature all-or-nothing". Hence, "statements of necessity and ineluctability will have to be reworked to incorporate a degree of probabilistic imprecision that better mirrors the exploratory state of the art of most of our sciences" (Rosen, 1982). It is this probabilism, the acceptance that "causality is a matter of degree", which opens up the possibility of analyzing social phenomena in terms of causal links apprehended through aggregate trends, regularities, and covariations. This makes a retreat into the micro world of the individual unnecessary and the critique of a collectivist methodology in social research unwarranted.

There are two more points of clarification which are necessary. First, it is essential to keep separate the two levels of analysis the micro and the macro. The fact that the latter is composed of micro events or conversely that the individual in his micro strategy will make references to macro patterns, does not mean that one level of reality can be understood in terms of the other, although they can shed light on each other. Let us consider for a moment the micro-exchange model of explanation based essentially on the calculus of exchange, whether the stress is on the cost-reward angle and hence behaviourist reinforcement or avoidance (Homans), or on mutual and reciprocal free relations of individuals (Gould), or on the rational 'bounded' choice of individuals in the situation of interaction (Boudon). As an example, let us assume that the 'exchange' is between two individuals who are engaged in trading ethnic prejudices. Now, there is no denying that in their strategies they will take into account, say, how widespread ethnic prejudices are in their pluralist society and what recources, power, and status are at the command of each ethnic group. Of course, the extent of information or ignorance regarding such macro patterns will affect the actors' actions, but the latter can nonetheless be adequately analyzed by the exchange model in view of the 'givens' of the situation which includes the measure of ignorance, the misconceptions of individuals, etc. However, we cannot see how this model can provide us with any knowledge concerning such macro patterns in society. Yet, it is these patterns that the sociologist wishes to apprehend and analyze and he cannot do so without resorting to aggregated and categorized data.

The second point is that since analysis of such data will basically take on the form of inferences of a probabilistic statistical nature, the 'human' criterion which is of significance at the individual level and is thought to destroy simple spatio-temporality, and hence the applicability of the causal principle, loses its relevance at the aggregate-holistic level. Furthermore, explanations based on such statisticalist causal inferences can be defended as scientifically acceptable. To quote just one philosopher of science, among many, who support the plausibility of the statisticalist approach: "This probabilistic construal of the explanation does not deprive it of its explanatory power. It is still a covering-law type of explanation, but the law is statistical rather than universal" (Salmon, 1971). But then can the probabilistic type of explanation be causal in the same sense as the universalistic type? (Rescher, 1970). If our definition of causality is the classical determinist one of events being absolutely necessitated in the manner of 'If A then always B', probabilistic explanation is not causal. However, on the indeterministic view, which relates to many areas of inquiry in which A itself may be variable in its effects so that only the *probability* of it producing B is observable in the way of statistical regularities, the probabilistic explanation that results from such data *can* be accepted as causal explanation both in its 'productive' and 'nomological' senses. We are inclined to argue on pragmatic grounds that since the social research area is primarily concerned with regularities of a probabilistic nature, which lend themselves to plausible causal inferences, the notion of causality be dislodged from its deterministic moorings and a wider conceptualization of causality be adopted so as to include human events which in the aggregate follow stochastic laws.

Conclusions

The foregoing treatment of the subject of causality in relation to social phenomena shows not only the paucity of law-like generalizations and the limitations encountered in theoretical explanations, but also the disagreements extant in regard to the overall scientific framework to be adopted. In other words, there is a state of uncertainty concerning paradigmatic orientations in the discipline.

Essentially, paradigms are the very broad outlines or basic principles upon which a discipline develops its specific theories. In sociology these outlines have been provided by the 'grand theories' and wide 'conceptual frameworks', or 'overall models'. These are paradigmatic in that they set the standards for the solution of the puzzles encountered in the discipline. In this sense sociology has not differed a great deal from the natural sciences. It has produced 'hard core' theories which have become entrenched as paradigms, such as functionalism. This process has resulted in theoretical and empirical growth and the paradigms themselves have undergone crisis periods, refutations, and modifications (Lakatos and Musgrave, 1970; Kuhn, 1970; Dixon, 1973).

None of the contending paradigms, however, has shifted the discipline decidedly in one direction, and this may account for the contradictory theoretical explanations put forward in such areas as social mobility, race relations, or deviance, to mention just a few examples. An undecided paradigmatic state could also account for the inability of the dicipline to produce higher level 'law-like' theoretical generalizations enjoying wide acceptance. It seems to us that these uncertainties and challenges will continue to hover over the discipline. Questions such as: what interpretation of the natural science model is most suitable for sociological inquiry (if it is to be used at all); or which 'grand theory' best serves the discipline, cannot be resolved easily. There is a crucial test, however, that can be applied at the empirical research level. This is the measure of success of the paradigm and methodology adopted in the quest for causal explanations of social phenomena.

Bibliography

Abell, Peter, 1971: *Model Building in Sociology*, Weidenfeld & Nicolson, London, p. 176.

Andreski, Stanislav, 1984: *Max Weber's Insights and Errors*, Routledge and Kegan Paul, London, pp. 46f, 110.

Bartos, O.J., 1967: *Simple Models of Group Behavior*, Columbia University Press, New York.

Balandier, Georges, 1975: *Gurvitch*, Basil Blackwell, Oxford, p. 81.
Bartholomew, D.J., 1973: *Stochastic Models for Social Processes*, Wiley, New York, pp. 7, 38.
Berger, Peter, L., 1969: *The Social Reality of Religion*, Faber and Faber, London.
Blalock, Hubert M., 1964: *Causal Inference in Non-experimental Research*, The University of North Carolina Press, Chapel Hill, pp. 9, 16.
Blalock, H.M. ed., 1971: *Causal Models in the Social Sciences*, Aldine, Chicago.
Blalock, Hubert M., 1976: 'Implicit Theories Underlying Macro-Level Analysis: Measurement Errors and Omitted Variables' in Thorns, David C. ed. *New Directions in Sociology*, David & Charles, London, p. 77.
Blalock, Hubert M., 1984: *Basic Dilemmas in the Social Sciences*, Sage Publications, Beverly Hills, p. 29.
Blau, O. & Duncan O.D., 1967: *The American Occupational Structure*, John Wiley, New York.
Boudon, Raymond, 1983: "Individual action and social change: a no-theory of social change", *The British Journal of Sociology*, Vol. 34, No. 1 (March, 1983): 118.
Brodbeck, May ed., 1968: *Readings in the Philosophy of the Social Sciences*, CollierMacmillan, London, p. 584.
Brown, Robert, 1963: *Explanation in Social Science*, Routledge and Kegan Paul, London, pp. 183-185.
Brown, Robert, 1976: *Rules and Laws in Sociology*, Routledge & Kegan Paul, London, p. 64.
Bunge, Mario, 1963: *The Myth of Simplicity – Problems of Scientific Philosophy*, Prentice Hall, Englewood Cliffs, N.J., pp. 182, 195, 203.
Catton, William R., 1976: 'Sociology for a Post-Exuberant World' in Thorns, David C. ed. *New Directions in Sociology*, David & Charles, London, pp. 36ff.
Cohen, L.J., 1962: *Diversity of Meaning*, Methuen, London.
Coleman, James, S., 1973: "Theoretical Bases for Parameters of Stochastic Processes" in Mapes, R.E.A. ed., *Stochastic Processes in Sociology*, The Sociological Review Monograph 19, University of Keele, July 1973.
Dixon, Keith, 1973: *Sociological Theory – Pretence and Possibility*, Routledge & Kegan Paul, London, pp. 23, 25, 29, 66.
Durkheim, Emile, 1938: *The Rules of Sociological Method*, University of Chicago Press, Chicago, p. 3.
Doby, John T., 1969: 'Logic and Levels of Scientific Explanation' in Borgatta, Edgar F. ed., *Sociological Methodology*, Jossey-Bass, San Francisco, p. 139.
Duncan, Otis Dudley, 1969: 'Contingencies in Constructing Causal Models' in Borgatta, Edgar F. ed., *Sociological Methodology*, Jossey-Bass, San Francisco, pp. 74-76.
Duncan, Otis Dudley, 1971: "Path Analysis: Sociological Examples" in H.M. Blalock ed., *Causal Models in the Social Sciences*, Aldine, Chicago, pp. 115-119, 123, 124.
Duncan, Otis Dudley, 1975: *Introduction to Structural Equation Models*, Academic Press, New York.
Emmet., D. & MacIntyre, A., eds., 1970: *Sociological Theory and Philosophical Analysis*, Macmillan, London, p. xviii.
Erikson, Kai T., 1973: 'Sociology and the Historical Perspective' in Drake, Michael ed., *Applied Historical Studies*, Methuen/The Open University Press, London, p. 16.
Galtung, Johan, 1967: *Theory and Methods of Social Research*, Universitetsforlaget, Oslo, p. 423.
Gellner, Ernest, 1964: *A Dictionary of the Social Sciences*, J. Gould & W.L. Kolb, eds., UNESCO/Tavistock, London, p. 435.
Goldthorpe, John H., 1980: *Social Mobility and Class Structure in Modern Britain*, Clarendon Press, Oxford, pp. 274-6.
Gould, Carol C., 1983: "Beyond causality in the social sciences: reciprocity as a model of non-exploitative social relations" in R.S. Cohen and M.M. Wartofsky, eds., *Epistemology,*

Methodology and the Social Sciences, Reidel, Dordrecht, Holland and Boston, U.S.A., pp. 53ff.
Gratch, Haya (ed)., 1973: *Twenty-Five Years of Social Research in Israel*, Jerusalem Academic Press, Jerusalem, p. 36.
Guttman, Louis, 1978: 'Recent Structural Laws of Human Behavior', Paper read at the Ninth International Sociological Congress in Uppsala, Sweden, August 14-19, 1978.
Hanson, N.R., 1965: *Patterns of Discovery*, Cambridge University Press, Cambridge, Chapter 3.
Harré, Rom, 1976: 'The Constructive Role of Models' in Collins, Lyndhurst, ed. *The Use of Models in the Social Sciences*, Tavistock Publications, London, p. 24.
Harré, R., and Secord, P.F., 1972: *The Explanation of Social Behaviour*, Basil Blackwell, Oxford, pp. 67, 73-76.
Hempel, Carl, 1965: *Aspects of Scientific Explanation*, Free Press, Glencoe, Ill. p. 243.
Hesse, Mary, 1976: 'Models Versus Paradigms in the Natural Sciences' in Collins, Lyndhurst, ed. *The Use of Models in the Social Sciences*, Tavistock Publications, London, pp. 4f.
Homans, George, C., 1982: "The Present State of Sociological Theory" *The Sociological Quarterly*, Vol. 23, No. 3 (Summer, 1982): pp. 285-299.
Hospers, John, 1956: *An Introduction to Philosophical Analysis*, Routledge & Kegan Paul, London, pp. 223-249.
Hughes, John A., 1976: *Sociological Analysis: Methods of Discovery*, Nelson, London.
Kemeny, Jim, 1976: 'Perspectives of the Micro-Macro Distinctions' in *The Sociological Review*, Vol. 24, No. 4. Nov. 1976.
Kuhn, T.S., 1970: *The Structure of Scientific Revolutions*, University of Chicago Press, Chicago.
Lakatos, Imre, 1970: 'Falsification and the Methodology of Scientific Research Programmes'. In I. Lakatos & A. Musgrave, eds., *Criticism and the Growth of Knowledge*, Cambridge University Press, Cambridge, p. 133.
Lakatos, I. & Musgrave, A., 1970: *Criticism and the Growth of Knowledge*, Cambridge University Press, Cambridge.
Land, Kenneth C., 1969: 'Principles of Path Analysis' in Borgatta, Edgar, F. ed., *Sociological Methodology*, Jossey-Bass, San Francisco, p. 4.
Lave, Charles A. & March, James G., 1973: *An Introduction to Models in the Social Sciences*, Harper & Row, New York, pp. 4, 5, 102f.
Lessnoff, Michael, 1974: *The Structure of Social Science*, George Allen & Unwin, London, pp. 16f., & Chapter 3.
Levy S. & Guttman, L., 1974: *Values and Attitudes of Israeli High School Youth*, Israel Institute of Applied Social Research, Jerusalem.
MacIver, R.M., 1964: *Social Causation*, Harper Torchbooks, New York, p. 386.
Meehan, E.J., 1968: *Explanation in Social Science*, The Dorsey Press, Homewood, Ill., pp. 11, 19-25, 68f.
Mennell, Stephen, 1974: *Sociological Theory-Uses and Unities*, Nelson, London, pp. 67, 87.
Nagel, Ernest, 1961: *The Structure of Science*, Routledge & Kegan Paul, London, pp. 324-335.
Papineau, David, 1978: *For Science in the Social Sciences*, The Macmillan Press, London, p. 67.
Parsons, Talcott, 1954: *Essays in Sociological Theory*, Free Press, Clencoe, Ill, p. 13.
Pawson, Ray, 1978: 'Empiricist Explanatory Strategies: The Case of Causal Modelling' in *The Sociological Review*, Vol. 26, No. 3., August 1978, pp. 613-624, 628, 632f.
Popper, Karl R., 1961: *The Logic of Scientific Discovery*, Hutchinson, London Science Editions, New York, pp. 59-61, 161, 205, 420.
Popper, Karl R., 1976: 'The Logic of the Social Sciences' in *The Positivist Dispute in German Sociology*, Heinemann, London, p. 89f.
Redfield, Robert, 1947: "The Folk Society", *The American Journal of Sociology*, Vol. 52, No. 4, January 1947, pp. 293, 299.

Reichenbach, H., 1961: *The Rise of Scientific Philosophy*, University of California Press, Los Angeles.
Rescher, Nicholas, 1970: *Scientific Explanation*, The Free Press, New York, pp. 122-130.
Rescher, N. & Ürquhart, A., 1971: *Temporal Logic*, Springer Verlag, Wien/New York, pp. 162, 200.
Rosen, Deborah A., 1982: "A Critique of Deterministic Causality", in *The Philosophical Forum*, Vol. 14, No. 2. (Winter 1982-83): pp. 115, 118f.
Rosenthal, R. & Rosnow, R.L., 1975: *Primer of Methods for Behavioral Sciences*, Wiley, New York, p. 35.
Rudner, Richard S., 1966: *Philosophy of Social Science*, Prentice-Hall, Englewood Cliffs, N.J., pp. 25, 26.
Russell, Bertrand, 1954: Book Review, *Observer*, London, 5 April 1954.
Ryan, Alan, 1970: *The Philosophy of the Social Sciences*, Macmillan, London, pp. 46ff., 51.
Salmon, Wesley C., 1967: *The Foundations of Scientific Inference*, University of Pittsburgh Press, Pittsburgh.
Salmon, Wesley C., 1971: *Statistical Explanation and Statistical Relevance*, University of Pittsburgh Press, Pittsburgh, p. 6.
Shye, Samuel, 1978: 'On the Search for Laws in the Behavioral Sciences' in S. Shye, ed., *Theory Construction and Data Analysis in the Behavioral Sciences*, Jossey-Bass, San Francisco, p. 18.
Simon, H.A., 1968: *International Encyclopedia of the Social Sciences*. Sills, D.L. ed., Macmillan & Free Press, Vol. 2, pp. 353-355.
Smelser, Neil J., 1968: 'Sociology and the other Social Sciences', in Lazarsfeld, Paul, F. et al, eds., *The Uses of Sociology*, Weidenfeld & Nicolson, London, p. 14.
Stinchcombe, Arthur L., 1968: *Constructing Social Theories*, Harcourt, Brace & World, New York.
Strawson, P.F., 1970: 'Dissolving the Problem of Induction' in B.A. Brody, ed., *Readings in the Philosophy of Science*, Prentice Hall, Englewood Cliffs, N.J. pp. 590-6.
Turner, Ralph H., 1964: *The Social Context of Ambition*, Chandler, San Francisco, pp. 49, 52.
Turner, Stephen P., 1982: "On the relevance of statistical relevance theory", *Theory and Decision*, Vol. 14 (1982): 195-205.
Weber, Max, 1949: *The Methodology of the Social Sciences*, The Free Press, Glencoe, I11. pp. 91, 101, 102, 103.
Whyte, W.F., 1943: *Street Corner Society*, University of Chicago Press, Chicago, pp. 3-25.
Willer, David & Judith, 1973: *Systematic Empiricism: Critique of a Pseudoscience*, Prentice-Hall, Englewood Cliffs, N.J.

Chapter 4

The Time Factor

Time, that is a delimited part of it, is often used by the sociologist, together with such limiting factors as geographical or societal area and specific population groupings, to pin down the boundaries of his research; that is the frame within which he carries out his investigation. In a very important sense, however, the time factor poses more complex and almost intractable methodological problems, and the mere fixing of time periods for one's research does not avoid these problems. The twin characteristics of time, omnipresence and intransitivity, make it elusive and most difficult to handle. It is because of the time factor, more than anything else, that the serious problems of reactivity and contamination arise and that the obstacles in the way of comparability and generalizibility are met. This must lead to the recognition that sociological experimentation close to the natural science model, where time is mostly successfully neutralized, is well-nigh impossible.[1]

Ultimately, the limitations imposed on sociological science do not derive solely from the time factor. The latter combines with space and population, and is further compounded by specific societal structuring and the changing patterns of the interplay between social phenomena, to form what is usually labelled as the *socio-cultural context*. But this resultant socio-cultural context itself is far from being hermetically cut off from previous historical and socio-cultural periods, and has its relationship with other population groups and cultures of various periods and locations. Furthermore, its contacts and the socio-cultural diffusions affect the contemporary as well as the more distant future societies, structures, institutions and so forth. As Edmund Burke put it: "society is a partnership not only between those who are living, but between those who are living, those who are dead and those who are yet to be born" (Hoffman & Levack, 1949). It is similarly true that aspects of continuity, gradual change and development, as well as more drastic changes, innovations and abrupt discontinuities characterize the social world. Hence the most complex relationships are present in any sociological topic under examination, characterized on the one hand by time, location and population specificities, and at the same time affected by the wider relationships across

[1] But see below a note on time-related puzzles in the natural sciences, even in laboratory experimentation.

periods, cultures, societies and social groups. We have chosen for the present discussion the time factor, which is only one element, albeit an important and far too often neglected one, that affects the process of research and explanation. By focusing on this element we hope to illustrate the enormous, if not entirely insurmountable, obstacles that sociological science faces.

The nature of time: philosophical views

The questions of how to define time and what is its nature appear to be enigmatic and have led to a plurality of conceptions, including metaphysical notions of time, physicomathematical, biological and psychological conceptions, and the idea of sociocultural time. In a by now classic piece, P.A. Sorokin has analyzed the differences between the variety of existing time systems and has stressed, in particular, the quantitative nature of time in physical sciences as against the qualitative nature of sociocultural time. He did not deny, however, the relationship between the two systems and the complementarity that exists between the two, so far as their functional efficiency is concerned (Sorokin, 1943). We wish to pursue the relationship from a notional perspective, that is to say the implications that the most advanced time notions of philosophers of physics have for the science of society. It is certainly beyond our scope or expertise to consider this topic in any detail, as for example to compare Newtonian absolute time with time in terms of the Minkowski-Einstein theory of relativity. It is worth, however, prefacing our short discussion with an observation we shall make concerning one of the chief criteria of distinction between these two physical time conceptions. "The time of classical mechanics is continuous, infinitely divisible, uniformly flowing, purely quantitative and devoid of any qualities"; on the other hand, "the noncontinuous, 'atomic', 'epochal' character of the time of quanta [is] connected with its principle of indeterminacy" (Sorokin, 1943). Yet, as we shall argue below, the latter conception as expressed in the 'four-dimensional manifold' has not only given rise to controversies concerning change and motion, but it also has deterministic implications.

In the first instance it is relevant to consider the controversy concerning the spatialization of time and thereby the unification of time and space. It ought to be clear that both dimensions are parts of the *unity* of actual experience – my walking a few paces is taking place simultaneously in both time and space and in this sense it represents a process which unifies time and space. This still allows the separation of the two by abstraction and, logically, the time element could have a quality or qualities different from the space element with which in experience it becomes fused. Hence there arise the important questions of whether time flows, in what sense it can be divided into past, present and future, and whether motion and change are real and the choice of future action possible. We shall consider the notion of the 'four-di mensional space-time manifold' with the questions enumerated in mind. According to this physicalist notion, physical objects and occurrences (presumably excluding human interactions) lie tenselessly along 'world-lines' i.e., tracks in space-time. This means that the four-dimensional manifold is a timeless web of 'world-lines'. It does not mean, however, according to these philosphers of the manifold, that interaction, motion and change are not taking place. Such matters are accounted for by the relative orientations of world-lines and the world-lines intersecting at different points. But in this view the passage of time is a myth, hence to say that the flow of time is unidirectional is wrong and irrelevant. The idea of past, present and future as time scales through which things change is due to a time illusion. Namely, all things are given in the time-space manifold, and 'change' means simply different positions along world-lines and various orientations and intersections of world-lines. The ideas of motion and change arise due to changes which occur in our states of consciousness "as we berome successively aware of adjacent cross-sections of the four-dimensional world" (Hobson, 1923; Grünbaum, 1962; Smart, 1967).

This manifold philosophy appears to be surrealist and metaphysical. Furthermore, it entails a clearly dualist and determinist world-view. It is possible to fit into such a dualist scheme the subjectivist Bergsonian notion of a time that is grasped intuitively, and a distinction could then be made between a psychological time on the one hand and a physical time on the other. But even if we agree that the latter is not an intrinsic structure of the physical world or a necessary attribute of nature, but rather the creation of the scientists, a conventional order or an assumption he makes to facilitate an explanation of the universe (Gunnell, 1970), and that, therefore, the manifold theory of time applies only to the material world, we are still confronted with the question of the relationship between mind and matter. Whether we equate the mind with the brain as an information processor or attribute to the mind transcendental qualities (Hofstadter and Dennet, 1981), the manifold theory will necessarily be dualistic and, quite curiously, must hold that minds are not in the space-time manifold; since only when minds, in which processes of consciousness occur, are in a different and non-spatialized time can the time illusion, or notion of change, or what have you, occur (Mundle, 1967). The manifold theory on this score remains indecisive and accentuates rather than resolves the mind-matter problem.

Our second contention is that the manifold philosophy provides a deterministic world-view. It could be argued that this world-view is irrelevant so far as the social scientist is concerned, since it attempts an explanation of the material world rather than the human world. Whilst it is neat to insist that different levels of discourse should not be confused, there are reasons why the 'four-dimensional manifold' theory should not be sidetracked in this way by the social scientist. First, it would mean accepting an extreme dualist position, holding that the material world on the one hand and the human world on the other are altogether unrelated and do not affect one another. Second, social theory has already been influenced by concepts arising out of the manifold theory (see below: modern industrial society seen as dominated by the spatialization of time) The possible implications of such a theory for social science are, therefore, worth considering, albeit briefly. Thus, in our interpretation, this world-view is both mechanistic and deterministic since, using the 'illusory-time perspective' language, the theory asserts that what will happen is destined to happen; in a sense has happened, since it is there in a timeless manifold. This is tantamount to engaging in a circular argument, namely saying that once a thing has happened it is proof that it was going to happen because it has happened, which means that it was there all the time. But there are no logical grounds on which to assume that this is so. Surely, there is as much logic in saying that things are indeterminate, i.e., that there are different possibilities, options, choices, etc., and that there is no *necessity* for a thing to happen, just because it *can* happen (Ayres, 1968). This does not mean that things happen haphazardly, without pattern, and not within certain constraints. But the patterns and constraints can allow different combinations, concatanations and outcomes. Hence what will happen is *genuinely* unknown and not given.

The latter conclusion need not imply that prediction in social science is impossible but it is another pointer to the reason why it is difficult to attain in social forecasting more than modest levels of probability. More positively, our conclusion highlights the possibility of choice of action and planning for the future. Such planning is a vital element in the social researcher's enterprise to reconstruct the past, to clarify trends, to forecast the future and to plan change. This is what sociological methodology is about and it is to such ends that it should be directed. But in doing so we raise further time-related riddles, such as the relationship between past, present and future, the nature of memory traces, and the meaning of future images and anticipations, and so on. We would digress here unnecessarily if we were to consider the arguments around the fascinating questions of time-anisotropy, that is the temporal irreversibility of processes, events, happenings, etc.; or philosophize about the present, which some see as inscrutable and 'specious'[1]; or, again,

[1] C.D. Broad points to the following in attempting to define the 'specious present'. If one takes sensing to involve a process with a finite duration, then anything an observer can sense at any moment "stretches a little way back behind that moment". Thus one cannot draw an absolutely sharp line between sensing and remembering (Broad, 1923). Furthermore, the time span of immediate consciousness is an interval within which what is earlier may be distinguished from what is later though both are directly present to consciousness.

ask what are the links between the past and temporal leaps into the future by means of mental images of the future contained in contemplation, anticiaption, etc. The point we do wish to make, however, is that sociologists and other scientists have attempted to define the very concept of time highlighting the different time horizons that can be conceived and the relationship between them. Thus, John Gunnell who raises the radical possibility that "time is not a necessary category of experience" and that " 'time' is not the name of some entity", defines it as "a symbolic form for organizing our experience of change and conceiving the relation between things" (Gunnell, 1970). Similarly, Niklas Luhman defines time as "the interpretation of reality with regard to the differences between past and future" (Luhman, 1976).

Time and social reality

This kind of modern sociological approach to time explains why time has been given a variety of concrete expressions and why it varies with different cultures. Cosmic or chronological time, for instance, belongs in particular to Western culture and in modern industrial society it is becoming the predominant time concept. To paraphrase Gunnell on this point: the mode of social organization becomes adapted to quantitative time; advanced technology requires not only rational time accounting, but also clear time planning. The future is, thus, made to appear, and perhaps is, less problematical; society begins to lose its past historical character and the gap between the present and future narrows. Moreover, planned development attempts to defuturize the future. In sum, through the spatialization of time, that is through the centrality given to the measurement of lived time as a way of organizing human activity, society "begins to appear more like the four-dimensional world of the physicist" (Gunnell, 1970).

Modern social research, both its methodology and theory, appear to acquire great significance in this process of mediating "relations between past and future in a present". Thus, to exemplify, social systems may be seen as "nontemporal extensions of time. They make the time horizon of other actors available within one contemporary present" (Luhman, 1976). And methodology, for instance, in stipulating causal relationships between events and in designing stochastic models, attempts to reduce the complexity of social life not only by analyzing and explaining past phenomena but also by forecasting emerging ones. Sociological knowledge from this perspective seems, at least partly, to be geared to social engineering. However, a few instances highlighting the stumbling blocks of social research will show immediately why a good measure of uncertainty and indeterminateness prevails in social science, and is thus hardly likely to produce 'precision social engineering'.

The stumbling blocks essentially emanate from the interpenetrating nature of the time categories we establish and we use as the parameters of research. This tends to blur and sometimes distort our vision and understanding. Thus, the shifting Now along the time-axis (Grünbaum, 1962), if we accept this imagery, means in effect that we are for ever endistanced from our field of inquiry, having to deal either with the more immediate past, the more distant past or the various futures to come. Seen another way, we have to investigate past presents and future presents. One methodological difficulty is that of investigating an earlier 'present' at a later time. The distance from the past cannot be avoided by reproducing the initial or antecedent conditions (Apel, 1979). We must resort, therefore, to some form of retrospective technique which could easily be prone to memory errors. We may interpret such errors in a much deeper sense than just the possible failings of the individual human brain to record and reconstruct events accurately. We could invoke the effects of what Halbwachs has called "collective memory", namely that the social milieu, whether that of the family or religious group, "gives a persistent framework into which all detailed recall must fit and it very powerfully influences both the manner and the matter of recall". This means that whilst memory is not *of* the group, it is always *in* the group. In this sense the past social framework or reference point provide the *schemata* within which the apprehension of time and the imaginative reconstruction, which we call memory, takes place. As an example, the memory of early communal organization reported by immigrants, which will have been influenced by their historical social groups, may produce a different picture from the one that we might have been provided with by a contemporary sociologist studying their communal organization at the time of arrival. (Bartlett, 1932; Halbwachs, 1968).

But even when the study is undertaken at the time of the social event taking place, the major research techniques run into difficulties. The participant observer's perspective may have changed with time as he turns to his records, field notes and impressions, and may become redefined on re-reading his information for the purposes of analysis. The interviewer has a different problem: he receives from the respondent data which at "the moment of measurement may compress within it a huge span of time", the effects of which are difficult to calculate. It will suffice to say at this point that the social researcher must be aware of the temporal character of social events. Thus, experienced time affects the definition of clock time by social actors according to their social scenes. The elapse of time, and with it the change in the socio-cultural environment, must not be allowed to distort the data obtained by means of retrospective techniques (Cicourel, 1964; Vidich, 1955; Hyman, 1955; Hyman, 1972; Galtung, 1967).

At the other end of the continuum, where we attempt to study the future, we must remember with Merton the effects of 'self-fulfilling' and 'self-destroying' prophecies; the question thus arising as to whether system-isolation and conditional prediction are at all possible. There are rationales for approximations of experiments, in the absence of true experiments, in order to achieve conditional prediction (Apel, 1979). Such rationales involve randomization and 'elligibility' criteria for the assignment of individuals to treatment and control groups. The units of assignment may also be institutions, localities or regions, and groups of individuals. Such designs could be successful in evaluating experimentally the results of some development programme, and in using it for prediction with regard to wider application. Thus, a regression-discontinuity design could show the effect of a compensatory reading programme, i.e., an effect of the reading programme seen in children's scores on reading tests would show-up as a discontinuity in the regression line (Boruch, 1976). It must be stressed though that multiple treatment and control groups, necessary because of the possibility of "different sets of inputs as a variation of the experimental treatment" (Bernstein, 1976), could widen the range or our predictions based on the assessments of the various combinations. But this also highlights the complexity of the future and its openness.

The difficulties encountered in achieving a quasi-nomological behavioural science built on experiments and conditional predictions has driven 'critical theorists' to seek a complementary social science which would be "a kind of scientific mediation of the society's self-reflection as a communication-community in making" (Apel, 1979). This would call for dialogue and deliberation among the subjects of planning, regarding the ways and means of social praxis, resulting in a normative social science which would move potentiality into actuality. We shall mention towards the end what this approach would be like in operational terms.[1] Undoubtedly it reflects another attempt to deal with the uncertainties of the open future, with its cumulated effects and unforseen events.

We have seen how intricate the very concept of time is and have attempted to present the particular implications such a concept has for social science. We shall continue to explore certain developments in social research which are related to the time factor. Before doing this let us clarify the sociologist's view of time. The Kantian notion of time belonging to the basic categories of understanding developed by the mind's activities was given a sociological perspective by Durkheim, who stressed that these basic categories are social constructs and that there are cultural variations in concepts such as time and space which are related to other aspects of social life (Goody, 1968; Fraisse, 1968; Durkheim, 1954). The variability of the notion of time is further

[1] See: 'action research'.

reflected in a number of ways. Thus, anthropologists claim that some societies do not have the equivalent of the category 'time'. Again, psychologists point to the conditioning that human beings are exposed to, due to the repetition of their cycles of activity, so that individuals' temporal horizons may differ. Finally, sociologists stress that each social framework, be it family, profession, church, nation, and so forth, has its own specific way of perceiving time, so that we may speak of "the multiplicity of social times" (Gurvitch, 1958). The variability of the notion and perception of time does not mean, however, that time is not always an aspect of experience implicit, as Goody says, in thought and speech. This experience of time takes the two major forms of *sequence* and *duration*, the interlocking of these two defining the process of change (Goody, 1968; Fraisse, 1968). The time dimension enters our experience also as scientists attempting to analyze causal process and change. The question that we wish to pose now is how the sociologist, *qua* social scientist, handles the time dimension.

The time dimension in social research

The sociologist who seeks to carry out his research within an empiricist scientific framework is confronted with the three basic ingredients of the scientific enterprise, each one of which contains a time dimension. The three elements are: data gathering, controlled comparison, and causal explanation. These are not separate elements; they are linked in a scientific process that ultimately should lead to general explanatory schemes dealing with the dynamics of society in its wider context. As Randall Collins argues, there is a serious risk of stagnation should sociology be tied to micro-level analysis, where results are applicable only to a particular group at a specific time. Data, its analysis and conceptualization, must refer to and summarize human behaviour and social events over long periods of time and wide areas; "the long-term and large-scale network of interaction" must be sought out, that is to say, the macro-level of analysis must be adopted (Collins 1975). In other words, guided by the basic aim of science, we must seek to establish the sorts of relations between phenomena whose explanation holds true over wide areas and long periods of time (Kuznets, 1934). Time-series analysis will illustrate for us the difficulty of achieving this aim.

A time series indicates the variation of a phenomenon over a longer period of time, whether observations on a variable are taken continuously or at discrete but regular intervals of time (Whittle, 1968). We could take as an example "how the percentage of women with no church affiliation varies from period to period" (Hagenaars and Cobben 1978). One such time-series shows that the percentage of women with no church affiliation increases *continuously* in some periods, but shows clear *accelerations* in other periods. The interpretation one is led to is that certain events which took place in the various periods were affecting non-affiliation differently in the different periods. One should avoid, therefore, the fallacy of 'period-centrism', that is the risk of generalizing the results of analysis concerning a particular period and applying them to other periods. In plain words this means that we cannot generalize since what we have is an *historical* or *uncontrolled* time series. There is, of course, an intrinsic difficulty in sociology of passing from an historical to a controlled sequence. In the natural sciences "experimental control of measurable phenomena makes it unnecessary for observations to be identified by historical dates". It is true that chronological sequence is preserved, but it is of no significance, since conditions in the test-tube or the testing laboratory may be kept constant over the various dates or times when measurements are taken. This means that the time factor can be eliminated, except where it is purposely introduced as an independent variable. Widely valid relations between variables are, therefore, much more easily established then in sociology. In the latter the interpretation is based on an historical series in which changes are brought about by a specific group of factors in each period (Kuznets, 1937).

Again, it is important to note with P. Whittle that: "In contrast to much experimental data the consecutive observations of a time series are not in general independent, and the fascination of time series analysis lies in the utilization of observed dependence to deduce the way the value of a variable can be shown to be in part determined by the past values of the same variable, or of other variables" (Whittle, 1968). A case in point may be the birth rates in a country. Other things being equal, the number of births at an earlier period will determine the number of potential mothers and, therefore, the number of births at a later period. Of course things are not likely to be constant in the social environment, so that other variables, such as birth control methods or religious attitudes, will most likely influence the birth rates and thus the number of births. Despite the difficulties of introducing experimental conditions, attempts have been made at designing experimental methods for the social sciences some of which only simulate the experimental, whilst others come closer to such conditions. Donald Campbell sets out and explains these attempts succintly. It appears that the time factor is one of the chief stumbling blocks. For instance in the 'one-group pretest-posttest' design there are several extraneous variables which are left uncontrolled and whose effects will be confounded, therefore, with the possible effect of the independent variable being manipulated. Two of these revolve around the time factor, namely: *history* and *maturation*. The former concerns the specific events that have occurred during the time that elapsed between position A i.e., before the test took place and position B i.e., after the test has been applied. Maturation on the other hand covers the effects which are systematic with the passage of time, such as the biological or socio-psychological factors affecting respondents during the period that the test takes place. Even other extraneous variables like the effect of *testing* and *instrument decay* are partially time induced, for it naturally takes time for a person to become sensitized to the test and produce different results a second or third time, and it is only with the passage of time that observers or raters become fatigued, experienced, etc. But supposing that these extraneous variables are controlled, as where we adopt a 'pretest-posttest control group' design; it is still practically impossible to achieve *simultaneity* between the test and control groups under observation. And whatever sophistication is introduced into the design, the *unique internal 'session' history* of each group still remains a possible confounding factor. (Campbell, 1970) If the time factor does not totally invalidate social science experimentation it certainly weakens explanations based on such experimental research.

Although we cannot attain true experimental conditions, in which time is neutralized, the logic of experiments in the social sciences follows closely that in the natural sciences.[1] The laboratory is replaced by a school, hospital, or some similar setting; the independent variable is actively manipulated, the extraneous variables are controlled, the subjects are assigned to treatments at random, and the dependent variable is measured. In principle this should enable the social scientist to establish causal connections. But there are practical difficulties in meeting these criteria in the 'field' (Krausz and Miller, 1974), due to the relatively long time that many social processes take to unravel, and the wide areas and complex relationships involved. Hence the sociological inclination towards macro-level analysis aided by designs covering large areas and seeking cross-sectional representativeness, often of large heterogeneous populations; the tendency to carry out replication studies; and the quest for cross-societal comparability. The time factor again looms large in these research designs.

[1] Les us note here that laboratory experimentation in the natural sciences is not as perfect as it is believed to be, and that within that framework, too, time-related problems have appeared. Natural scientists have acknowledged certain fundamental weaknesses in their laboratory experiments and have pointed to many erroneous claims made on the basis of scientific experimentation which were later discredited. An example is Michael Polanyi in whose view "It is always conceivable that reproducibility depends on the presence of an unknown and uncontrollable factor which comes and goes *in periods of months or years* (our italics) and may vary from one place to another". He supports his view by describing his own experiences in chemical laboratory research, but mentions other fields, too, such as biology and physics. The underlying problem is that even the purest substances available contain traces of various chemical elements which could trigger-off 'epidemics'. Hence such substances "may reproducibly maintain a certain character for a time and then suddenly take on once more reproducibly for another period a different one, depending on *periodic variations* (our italics) of the elements present in traces ... They may come and go without the cause having been discovered. [Hence] there is always a *conceivable* doubt as to the convincing power of reproducibility" (Polanyi, 1946; Polanyi, 1958).

Let us first consider the cross-sectional study. This has been defined as "the collection and analysis of data, grouped according to age, of a 'cross-section' of the population at one particular moment". Effectively this means that different cohorts of respondents which could be regarded as sub-samples, are surveyed or tested simultaneously for comparative purpose. Does comparative evidence adduced in this way prove relationships between variables or causes of change? Supposing we investigated attitudes to political issues and we came up with a close correlation between more advanced age and political conservatism. It would be wrong to conclude that as we grow older we become more conservative, since the respondents in the different cohorts were born in different years and, therefore, they had different political experiences. For, we must not fall prey to the life-course fallacy, that is "interpreting differences between the various age groups at a certain moment as being purely the result of differences in age, [since] cohort effects also play a role". We must not confound age differences with the time of birth differences (Hagenaars & Cobben, 1978; Krausz & Miller, 1974). At this point let us define 'cohort' and clarify the use of cohort analysis. A cohort is "a group of people born during a particular time interval"; the use of cohorts enable us to undertake cohort analysis which involves "the examination of three kinds of effects or sources of change: those due to the life cycle (age effects); those due to generational change (cohort effects); and those due to influences associated with particular periods of time (period effects)" (Seidman, 1976). We shall give below further examples to illustrate the fundamental analytic difficulty in cohort analysis due to the likelihood of confounding the three effects mentioned.

But first let us consider whether it is possible to overcome such confounding by adopting a longitudinal design. The latter involves repeated measurements of the same samples or cohorts over time, and it has been asserted that it is the only method which can give "a true picture of cause and effect relatioships over time" (Wall & Williams, 1970). Longitudinal studies are beset, however, with serious weaknesses due precisely to the time factor. Thus sample attrition means that after a longer period comparability between the initial and final groups is impaired. Apart from this, as with any panel study, there are conditioning and testing effects as mentioned above, and there is also the possibility that some disturbing variable may have intruded between the recurrent measurements.

Longitudinal designs, which have been used to good effect in developmental studies connected with biological and psychological research and to a lesser extent in sociological analysis, appear to be concerned with two broad problems: (a) prediction, to be achieved by "monitoring the stability of defined characteristics of a defined population over time" and by "exploring how far such characteristics may be modified" through future intervention;

(b) genesis, to be understood by studying the stages, sequences and rates of development. But the difficulties connected with the longitudinal method have not been resolved, basically because of the inability to control the complex relationships of the many variables involved over the long periods that longitudinal studies take to be completed (Wall & Williams, 1970). Within the longitudinal framework different *sequential designs* have been proposed, therefore, so as to overcome the confounding of the various sources of change. For instance, in the cohort-sequential design two cohorts are studied simultaneously time and again, in order to separate out changes due to age and possible effects due to the specific times when measurements were being taken. The time-sequential design on the other hand attempts to separate out differences due to age from the effects of being born in one year rather than another. It does this by taking measurements at time intervals of two sets of cohorts with different years of birth. Further sophistication is introduced by a combination of the two approaches instanced, a design known as cross-sequential (Helmstadter 1970). The aim of these complicated designs is to study changes and developments in human capacity, human behaviour, or social phenomena over time, ensuring though that the different aspects of the time element are not confounded with various other sources of change.

The time element is central also to replication work. In contrast to longitudinal studies which seek to uncover trends, developments and causes of change, replication work is essentially concerned with the falsification or verification of theoretical propositions arising out of previously executed research. Strict comparability is therefore desirable, a requirement which necessitates that the replication studies be carried out within the same *era* or *period* as the original study. Herbert Hyman focuses on the concept of an era or period and in a somewhat different context considers its implications. Following his ideas we may develop the following questions that appear relevant to our topic of discussion: First, how important in the causative sense is "the spirit of the times and its web of institutions" as opposed to the single event which is being proposed as a causative agent? That is to say, can one really separate our effects of single events from the holistic effects of entire eras? Second, can one regard the holistic effects so dominant as to regard any internal variations revealed as reflecting merely some measure of instability or the unreliability of the methods employed, and, therefore, are we to evaluate the variation *between* periods, on the basis of holistic assumptions? Third, what are the effects of former eras, that is to say what residues of previous social processes and institutions will be found in the later eras? Fourth, how is time to be segmented; namely, when does an era start, when does it come to maturity and what are the signs of it having ended? Do eras start and end abruptly or do we have to view them as having more elasticity, that is "not as separated by sharp boundary lines but rather by border zones of time, when there is a gradual transition and emergence of new forces"? (Hyman, 1972).

Time and causal sequence

Much of our discussion has revolved around the problem of our explaining sequential changes in social phenomena and of understanding the particular relationships between phenomena within some time context. Since 'the present' has no definable extent (St. Augustine quoted in Hyman, 1972) and can be philosophically conceived only in terms of an infinitesimally minute unit of time which is unknown, or better still a durationless moment if such is imaginable, all social phenomena are perceived as unfolding within some known, definable and measureable time period, which are referred to as 'the past', 'the future' and, instead of the inconceivable 'present', as 'the contemporary', in order to provide us with recognizable lables of the time dimensions involved. Now, where relations between phenomena are invariant and succeed in time, they signify causality. We may stress with Timasheff, therefore, that "causality is applicable only if phenomena are contemplated in the time dimension" and it does not denote simply asymmetrical relationship or some kind of functional relationship, in which cases temporal sequence and causality are not necessarily integral parts of the explanatory scheme. What matters, therefore, are the facts of sequence and duration, as mentioned above, and our ability of thus comprehending causal relations and change. This means that identifying static order, reflected in the coexistence of phenomena, is insufficient for revealing causality. As an example, the correlation between a low birth rate and a high degree of industrialization is no proof in itself of a causal connection and it certainly does not reveal the direction of causality. Only evidence of the trend over time by means of series of correlations at different periods may show how the introduction of industrial conditions has brought in its wake a lowering of the birth rate. This type of causal sequence reflects dynamic order *along* one causal chain.

Supposing, however, that we are dealing with a phenomenon produced by multiple causal factors through the intersection of two or more causal chains. An example might be the holocaust endured by European Jewry during World War II, which could be regarded as the outcome of a causal chain stretching back to the Jewish people's dispersal from its original homeland followed by its settlement in basically hostile antisemitic environments and its inability to organize its own defence, being intersected by another causal chain in which modern Germany's economic and political ills were successfully exploited by the ruthless Nazi movement bent on dominating Europe and exterminating the Jews. As Timasheff points out, although "a theoretical proposition is characterised by absence of reference to concrete time and space", such an intersection of causally relevant developments does take place in concrete time and space. He calls this intersection of independent causal development *conjuncture* and holds that whilst such developments are explainable after the events, they are not predictable (Timasheff 1959). That is, we cannot foresee the presence or absence, and the specific intersection points of the various factors at the particular times when, and places where they occur. Here we see clearly the limitation imposed on the feasibility of sociological explanation and prediction by the time and space factors. The difficulty of predicting social events is a particularly serious constraint for the applied sociologist involved in planning or intervention research. However, we do give some examples below of techniques which can enhance, even though to a limited extent, our predictive powers.

The effect of time on research techniques and analysis

We have pointed to the ways in which the explanatory and predictive powers of sociological research are weakened by the difficulty of contending with the time element involved. We highlighted these problems by considering some of the basic research designs usually resorted to. We have not paid attention to the time factor as it relates to the actual research techniques, whether this be sampling practice, questionnaire construction and use, or statistical data analysis. As our brief illustration will indicate, insufficient account of the time factor in the use of techniques weakens the validity of findings. Thus, sociologists are accustomed to using space or population sampling, and where applicable document sampling. Time sampling, in which the universe is any relevant period of time, divided up into time units from which the random selection can be made, is infrequently used. An example might be the observation over such time units of certain activities, such as those of children in a hospital or of car drivers on roads. But it is even more rare for time sampling to be combined with, say, population sampling, despite its distinct advantage of cancelling out 'time of day' or 'season' effects on interviews or observation, and thereby strengthening the validity of findings (Krausz & Miller, 1974). In questionnaire construction too, the location of a particular question (which provides the indicator for a variable) in the overall questionnaire context (Hyman, 1972) and the 'timing' of its administration to the respondent could have their effects, not only in terms of the way the sequence of the questions possibly affect their meaning and the responses to them, but also in terms of whether the question is responded to early on or later in the interview or 'filling-in' process. To randomize the location of questions, so as to overcome these problems, present technical difficulties and it may affect adverseley 'the flow' of the interview and the better understanding of the questions by the respondent due to the logical sequence inherent in the substantive aspects of the topic investigated. The time factor, therefore, once again tends to be neglected. At the stage of data analysis there are techniques which will eliminate the time factor as, for instance, through 'trend correction'. This is done by substituting the overall trend, say in marriage rate, found over the entire time span covering the whole research population, for the specific values in, say, particular periods of time. The remaining differences between the cohorts are then regarded as simply cohort effects. But since there is a complex interaction between cohort and time or period effects, it will be necessary to assess statistically the interaction that exists between these elements (Hagenaars & Cobben 1978).

The importance of the time element and its complex nature is revealed especially where several interrelated explanatory variables are used in an analysis. A cohort-analysis of political party identification and support, undertaken by Philip Converse, is most instructive in this respect. The basic question is how to explain the greater conservatism and higher degree of partisanship of older respondents. Do generational differences or maturational factors produce such a correlation? Converse suggests that plausible as such explanations may appear to be, they do not really explain the complex relationships of life-cycle, period, birth-cohort and time influences, and puts forward the mini-theory that "identifications intensify as a function not of age per se, but rather as a function of the length of time that the individual has felt some generalized preference for the particular party and has repetitively voted for it". The complexity of the situation, so far as party preferences are concerned, is seen in that cohorts show internal variations over time, and that, as Converse warns, the same historical configuration could have different effects on various age groups and that the flow of events some time later could have a totally different configuration of impacts (Converse, 1976).

The possibility of providing time-oriented explanations is explored by Bobrow and Cutler in a study of national security beliefs of the American public. Their purpose is to examine the extent to which three explanatory concepts: cohort, life-stage, and situation account for similarities or differences of beliefs. That is to say, how is belief associated with membership of a particular cohort, occupancy of a particular life-stage, or the situation prevailing at the time the belief was verbalized. A time perspective is involved in all three of these explanatory concepts- First since each person is a member of a particular birth-cohort, i.e., although all those born into a birth-cohort are not necessarily or strictly coeval, but they are "born during the same time interval and age together" (Ryder, 1965), the set of historical and social circumstances which they have an opportunity to experience is the same. Secondly, Bobrow and Cutler stress that each person passes through a series of life stages, occuppancy of different life stages taking place in an irreversible sequence i.e., childhood to adolescence, to adulthood, etc., this being terminated only by death, and that this occupancy of a life-stage is strongly associated with the age-groups a person is in, a factor determined by the particular culture in which he lives. Thirdly, these two properties do help us understand people's activities, but they are always observed in the context of some situation. Therefore, "it is always possible that what we see when we observe people at a particular point in time can be accounted for by the situation rather than life-stage or cohort differences" (Bobrow & Cutler, 1967). The particular point in time may, thus, be of prime importance.

The sorting out of these explanatory variables will have important repercussions on our ability to undertake social forecasting. Our power of predicting the long-term effects of social policies for instance will be enhanced by a refined analysis of cohort differences, and by our study of the effects of historical experiences especially in terms of "the time interval or lag between the experiencing of an historical situation and the point when its impact alters" (Bobrow & Cutler 1967).

The two studies we have quoted from show an awareness by some sociologists and other social scientists of the need to take into account the time element. Yet this may be the exception rather than the rule. Certainly this is the view of Tuma, Hannan and Groeneveld who state:

> "Though most sociologists profess an interest in social processes – how social behavior and social systems change over time – this interest is seldom reflected in their research. They usually examine relationships among phenomena at only one point in time. Even when temporal data are used, as in experiments and panel studies, sociologists seldom study dynamics – the time paths of change. Instead, the focus is on change from one equilibrium level to another, as measured, for example, by pre and post experiment differences or by levels at successive waves of the panel" (Tuma, Hannan, & Groeneveld, 1979).

In order to remedy the situation Tuma and colleagues wish to stimulate an interest in the collection and analysis of 'event histories'. An event history "records dates of events that occur for some unit of analysis", examples being "changes in categorical variables describing individuals such as marital status, employment status, health status and membership in voluntary associations, as well as those applying to social collectivities, such as changes in political regimes and outbreaks of strikes, riots, and wars." In an empirical illustration the researchers test the effects of an income maintenance experiment on change in marital status. What is to be noted in particular regarding their study is that they not only deal with population heterogeneity by reporting on the effects of experimental treatments on rates of marriage or marital dissolution, but they also report on time dependence by reporting on how treatment effects vary over the experimental period. They use a continuous-time stochastic model in order to describe the state of the categorical variable at every moment in time. Conceptually such a continuous model is acceptable although most social data fit better discrete models. Thus, one can imagine people having opinions as to how to vote at every moment in time, but what counts is the opinion they have at certain crucial times such as at elections. One can understand, however, the authors' wish to utilize all the information, that is the number, sequence, and timing of events. This, they claim, is superior to the conventional techniques, as for instance analyzing cross-sectional data, since social scientists wedded to cross-sectional analysis "seldom seem to reflect on the likelihood that equilibrium exists or on the length of time required for a system to reach a new equilibrium following some intervention or structural upheaval" (Tuma et.al., 1979). It seems very convincing to argue therefore, with a growing number of social scientists, that event-counts, event histories, and the study of transition processes and time dependence are necessary if causal analysis and prediction is to make a real advance in the field of social research (Tuma, et.al., 1979, Singer and Spilerman, 1976)

Divergent views of the time-variable

Although the techniques adopted in dealing with the time factor vary and are often inadequate, most social scientists, perhaps all, recognize the problem that this factor produces where the explanation of social reality is attempted. Even so, not only the techniques, but also the philosophical angles from which this problem is seen differ widely.

Let us juxtapose the views of Georges Gurvitch and Julian Simon by first quoting from and then somewhat elaborating on their writings:

> "... social life always takes place in divergent and often contradictory manifestations of social time. Difficult as it may be, every society must attempt to unify, even if only relatively, these multiple manifestations of time and attempt to arrange them in a hierarchy. This is unavoidable since each partial element of the global or all-inclusive society tends to move in its own time .... A society cannot survive without some unification of the plurality of its social times, yet special problems emerge in the very effort to achieve some consistency in the matter; ... diverse social times interpenetrate one another and are unified in a particular way. But how can these interpenetrations and the unification of these times corresponding to concrete collectivities be *studied* (our italics) without first establishing a difference between them and following their complex dialectics?" (Gurvitch, 1964).
>
> "We say that changes occur with time. True enough. But it is not time

itself that brings the changes to pass: rather various processes that occur over time bring change. If people were frozen in such a way as to slow all their bodily processes but not to kill them as can be done with small cells and human sperm the passage of time would have no effect. Rather, it is *fatigue* that affects the chimpanzee's performance late in the day. It might be growing annoyance that produces a different answer to a question late in an interview. It is changes in the world situation that alter the election-poll results over time. And it was a shift in tastes as well as improvements in tobacco and cigarette-making technology that caused the increase in United States cigarette consumption. The word 'time' itself is nothing but shorthand for other changes that occur." (Simon, 1969).

The divergence of views between these two social scientists is clear: Gurvitch has a relativistic approach to time, on the lines of the multiplicity of times demonstrated in quantum physics, in which not only are there "differences of measures applied to diverse times but also ... qualitative differences and varied relations of past, present and future in these times" (Gurvitch, 1964). Similarly with social time where the relativistic aspect is seen, as Maurice Halbwachs has pointed out, in the clash between the time dimensions of the family, school, factory, union, administrative office, etc. The complexity of social time is also due to the reciprocal relationship between social frameworks and time. Thus social phenomena condition and produce time but are also conditioned by and are the products of time; they give bulk to time but also elapse in time. As a result of this relationship Gurvitch concludes that "social time cannot be defined without defining the total social phenomenon" which can only be achieved through sociological conceptualization, and that the complex character of the social manifestations of time can only be described and studied by means of an operational framework to be provided by sociological theory (Gurvitch, 1964). Gurvitch propounded a 'depth sociology' which through a dialectical approach could uncover the many social layers of society in order to penetrate the superstructures of social reality and arrive at the more basic social patterns (Friedrichs, 1970; Maus, 1971). His notion of social time fits into this scheme in that "social life always takes place in divergent and often contradictory manifestations of social time". The social character of time is further reinforced by his observatrion that "time which has been unfolding, which is in a sense completed, has very few traits in common with time in the process of being made" (Gurvitch, 1964).

The stance taken by Julian Simon, on the other hand, seems to be quite different. Time for him is not a social product, it is not socially made but it is rather a given neutral datum. Time, perhaps like the existence of the universe, is non-social, a neutral framework within which social phenomena unfold, seen in their changes, developments and causal relationships. Therefore the *time-variable* is only a proxy for other processes and changes that operate over time. Looking at things in a cause-effect manner we may find that it takes a long time for an independent variable to have its effect on the dependent variable, like syphilis contracted by a young man causing him to go mad many years later, or the unpleasant remarks of a parent-in-law causing a breakdown in the marriage of a young couple some years later. But Simon insists that in such examples of long "latency period" between cause and effect, "it is not *time itself*" that produces the delayed effects but the series of "changes that take place unobserved and in interaction with subsequent happenings". Simon agrees that there are obstacles in the study of changes over time. This is so, for instance, where a comparison of several stages of development or of change is attempted, but where this is bedevilled by what he calls "time-related breaches of *ceteris paribus*". Thus, if we wish to find out what are the effects of advertising over a period of time, the change in consumption observed may be a result of changes in the socio-economic circumstances prevailing and not only of the cumulative effects of the advertising itself. Other things do not remain equal over a period of time. Julian Simon offers a number of solutions which come some way towards mitigating the obstacles. First, the researcher must separate the changes caused by the variable in question from changes due to other extraneous variables. He is to do this by working with appropriately chosen time periods. The effect of advertising must be measured as between comparable time periods i.e. in the same seasons, during periods with similar socio-economic conditions, etc. Second, the difficulty of measuring change over very long periods of time may be overcome by speeding up time artificially. Thus, making certain assumptions about the typical behaviour of individuals and institutions in the economic sphere, one can 'simulate' their interaction and run through the computer, in minutes, a number of business cycles that in reality will have taken many years. The economist can, thus, uncover the sort of business cycle that would result in the economy given a certain basic set of assumptions (Simon, 1969). Computer simulation could conceivably also be used by the demographer or sociologist to enhance his predictive power concerning future demographic trends in certain assumed conditions, or the rate of school integration given some basic assumptions relating to an ethnically plural society.

Conclusions

Now let us pause for a moment and summarize the major time-related questions in the realm of social research. Briefly they are: 1) temporal variablility; 2) reactivity with the passage of time; and 3) the time cycle effects of the research process itself.

Temporal variability is reflected in the different meanings, contexts, and definitions of a particular time or period not only on the historical time scale as expressed in Karl Mannheim's conceptualization of a 'generation', characterized by a set of people occupying a particular "location in the historical dimension of the social process" (Mannheim, 1952), but also contemporaneously as expressed in Gurvitch's 'social time'. Temporal variability raises the central issue, as to whether a greater stress on comparative diachronic research, favouring studies that cut further back in time and utilize unexploited historical material, should be contemplated by the sociologist. Some would call 'present-minded' research anything conducted within a 'generation' or a life-span of some sixty-years (Thrupp, 1970). The basic question is to what extent such historical sociology going well beyond the recent past will have to heed historicism, which sets itself against generalizations. But since any well-grounded sociology must be historically informed, does the time variability involved prevent it from producing "a transhistorical sociological science"? (Collins, 1975). Indeed, can sociology arrive at generalizations which transcend time and space specificities, thus resulting in *the laws* of this social science?[1] Reactivity over time produces different kinds of uncertainty. Reactivity may be seen as the effects of social phenomena, to be gauged by the reactions to the social phenomena being experienced. Arthur Stinchcombe maintains that the effects of, and therefore presumably the reactions to, certain social forces will be proportional to the "temporal 'size' of a group" i.e. "the time spent in different attention structures". Thus, the more time children spend in the street, the greater the subcultural influences of street-corner gangs; the more time spent in voluntary association meetings and activities the more intense will become the commitmens to the common values of the association; and so on. Therefore, "those *groups* which occupy large time-segments will have more causal influence on the culture of their members" (Stinchcombe, 1968).

Another possible pattern that emerges in connection with reactions to events or past trends is one which is reflected in the paradox of predictions. As Robert Friedrichs says: the very knowledge of social regularities offered by the sociologist acts to free the individual or the group from their re-enactment. Through 'feedback', 'self-fulfilling' and 'self-defeating' prophecies, regularities may be broken and the future may be freed from the past (Friedrichs, 1970). This role played by the sociologist of bringing about change in social patterns is expressed more clearly in the applied field where research is sociotherapeutic in nature. Such an approach, often called 'action research', aims at facilitating change by involving the subjects of the research

[1] In Chapter 3 we referred to attempts to establish such laws.

process itself such as the involvement of delinquent boys in research experiments around a club programme which is intended to reduce or eliminate their delinquent activities. Whatever the strengths or weaknesses of such research from a purely scientific point of view, it represents most clearly the interaction over time between the research process itself and the subjects of investigation.

Finally, in any social-scientific research, the very fact that research takes time and the constraints produced by the cycle of events in the process itself could affect its validity and applicability. Precautions notwithstanding, the early stages of a research programme could appreciably affect the later stages. The subtle influence on the researcher through gaining experience in the field and the possibility of 'contamination' of subjects are examples of how the research process may influence and thus invalidate the results. Furthermore, independent changes in the situation during the course of the research could label the findings as 'out-of-date' and, therefore, inapplicable. Such operationalist dilemmas need not deter us from building a social science, any more than the philosophical observation that "scientific theories are historically situated too, changing over the course of time" (Collins, 1975), so long as we take into account the difficulties posed by the dynamic nature of social processes unfolding in time.

Bibliography

Apel, Karl Otto, 1979: "Types of Social Science in the Light of Human Cognitive Interests" in S.C. Brown, ed., *Philosophical Disputes in the Social Sciences*, Harvester Press, Sussex, pp. 21, 24.

Ayers, M.R., 1968: *The Refutation of Determinism*, Methuen, London.

Bartlett, F.C., 1932: *Remembering*, Cambridge University Press, Cambridge, p. 296.

Bernstein, Ilene N., 1976: "An Overview" in I.N. Bernstein ed., *Validity Issues in Evaluation Research*, Sage Publications, Beverly Hills, Calif., p. 11.

Bobrow, Davis B., & Cutler, Neal E., 1967: "Time Oriented Explanations of National Security Beliefs: Cohort, Life-Stage and Situation" in *Peace Research Society: Papers*, Vol. 8, The Hague Conference, pp. 32, 33.

Boruch, Robert F., 1976: "Coupling Randomized Experiements and Approximations to Experiments in Social Program Evaluation" in I.N. Bernstein ed., *Validity Issues in Evaluation Research*. Sage Publications, Beverly Hills, Calif., p. 47.

Broad, C.D., 1923: *Scientific Thought*, Kegan Paul, Trench, Trubner & Co. Ltd., London, pp. 348-351.

Campbell, Donald T., 1970: "Factors Relevant to the Validity of Experiments in Social Settings" in Denzin, Norman K., *Sociological Methods*, Aldine Publishing Co., Chicago, pp. 244 ff.

Cicourel, Aaron V., 1964: *Method and Measurement in Sociology*, The Free Press, Glencoe, Ill., pp. 27, 47, 117.

Collins, Randall, 1975: *Conflict Sociology: Toward an Explanatory Science*, Academic Press, New York, pp. 11, 23, 35.
Converse, Philip E., 1976: *The Dynamics of Party Support*, Vol. 35, Sage Library of Social Research, Sage Publications, Beverly Hills, Calif., pp. 12, 19.
Durkheim, Emile, 1954: *The Elementary Forms of the Religious Life*, Allen & Unwin, London.
Flew, Antony, 1979: *A Dictionary of Philosophy*, Pan Books, London.
Fraisse, Paul, 1968: "Time I: Psychological Aspects" in *International Encyclopedia of the Social Sciences*, ed. David L. Sills, Vol. 16, The MacMillan Co., & The Free Press, U.S.A., p. 25.
Friedrichs, Robert W., 1970: *A Sociology of Sociology*, The Free Press, New York, pp. 184-6; 188-190.
Galtung, Johan, 1967: *Theory and Methods of Social Research*, Allen & Unwin, London, pp. 475 ff.
Goody, Jack, 1968: "Time II: Social Organization" in *International Encyclopedia of the Social Sciences*, ed. David L. Sills, Vol. 16, The MacMillan Co., & The Free Press, U.S.A., pp. 30, 31.
Grünbaum, Adolf, 1962: "The Nature of Time" in Robert G. Colodny, ed., *Frontiers of Science and Philosophy*, University of Pittsburgh Press, Pittsburgh, pp. 149-183; 154.
Gunnell, John G., 1970: "Development, Social Change and Time", in Dwight Waldo, ed., *Temporal Dimensions of Development Administration*, Duke University Press, Durham, N. Carolina, pp. 80, 81, 83-87.
Gurvitch, Georges, 1958: "Structures Sociales et Multiplicité des Temps" in *Société Francaise de Philosophie, Bulletin* 52, pp. 99-142.
Gurvitch, Georges, 1964: *The Spectrum of Social Time*, D. Reidel, Dordrecht, pp. 13, 151; 24; 13, 37; 30.
Hagenaars, Jacques A., & Cobben, Nico, P., 1978: "Age, Cohort and Period: A General Model for the Analysis of Social Change" in *The Netherlands Journal of Sociology*, Vol. 14, No. 1, July, 1978, pp. 64; 59-60; 75-6.
Halbwachs, Maurice, 1968: *La Mémoire Collective*, Presses Universitaires de France, Paris, Chapters II and III.
Helmstadter, G.C., 1970: *Research Concepts in Human Behavior*, Appleton-Century-Crofts, New York, p. 836.
Hobson, E.W., 1923: *The Domain of Natural Science*, Cambridge University Press, Cambridge, Chapters VI and XIV.
Hoffman, Ross J.S., & Levack, Paul, eds., 1949: *Burke's Politics*, Knopf, New York, p. 318.
Hofstadter, Douglas & Dennett, Daniel C., 1981: *The Mind's I*, Harvester Press, Brighton, Sussex.
Hyman, Herbert H., 1972: *Secondary Analysis of Sample Surveys*, John Wiley, New York, pp. 12, 229, 230; 228; 216.
Hyman, H. et.al., 1955: *Survey Design and Analysis*, The Free Press, Glencoe, Ill., p. 194.
Jung, H.Y. & Jung, P., 1976: "Time, Consciousness and the Structure of Human Action", paper delivered at the Edinburgh IPSA Congress, August 16-21, 1976, International Political Science Association.
Krausz, Ernest & Miller, Stephen H., 1974: *Social Research Design*, Longman, London, pp. 93, 40, 42-3.
Kuznets, Simon, 1937: "Time Series" in *Encyclopedia of the Social Sciences*, ed. E.R.A. Seligman, The MacMillan Co., New York, pp. 629, 630.
Luhman, Niklas, 1976: "The Future Cannot Begin: Temporal Structures in Modern Society", in *Social Research*, Vol. 43, No. 1, Spring 1976, pp. 135, 137, 149.
Maus, Heinz, 1971: *A Short History of Sociology*, Routledge & Kegan Paul, London, p. 208.
Mannheim, Karl, 1952: "The Sociological Problem of Generations" in *Essays on the Sociology of Knowledge*, ed., P. Kecskemeti, OUP, New York, p. 290.

Mundle, C.W.K., 1967: "Consciousness of Time" in *The Encyclopedia of Philosophy*, Vol. 8, ed., P. Edwards, McMillan & Free Press, New York, pp. 136-138.

Polanyi, Michale, 1946: *Science, Faith and Society*, Geoffrey Cumberlege-Oxford University Press, London, pp. 78-80.

Polanyi, Michael, 1958: *Personal Knowledge: Towards a Post-Critical Philosophy*, Routledge & Kegan Paul, London, p. 293.

Ryder, Norman B., 1965: "The Cohort as a Concept in the Study of Social Change" in *The American Sociological Review*, Vol. 30, No. 6, December, 1965, p. 844.

Seidman, David, 1976: "Cohort Analysis" in *Social Indicator Newsletter*, No. 10, Center for Coordination of Research on Social Indicators, The Social Science Research Council, Washington, September, 1976, pp. 1-2.

Simon, Julian, L., 1969: *Basic Research Methods in Social Science*, Random House, New York, pp. 136-7, 149-152.

Singer, Burton & Spilerman, Seymour, 1974: "Social Mobility Models for Heterogeneous Populations" in H.L. Costner ed., *Sociological Methodology: 1973-1974*, Jossey-Bass, San Francisco.

Sorokin, Pitirim, A., 1943: *Sociocultural Causality, Space, Time*, Duke University Press, Durham, N. Carolina, Chapter IV; pp. 159, 160.

Smart, J.J.C., 1967: "Time" in *The Encyclopedia of Philosophy*, Vol. 8, ed. P. Edwards, McMillan & Free Press, New York, pp. 126-132.

Stinchcombe, Arthur, L., 1968: *Constructing Social Theories*, Harcourt, Brace & World, New York, pp. 240-242.

Thrupp, Sylvia, L., 1970: "Diachronic Methods in Comparative Politics" in R.T. Holt and J.E. Turner, eds., *The Methodology of Comaparative Research*, The Free Press, New York, p. 346.

Timasheff, N.S., 1959: "Order, Causality, Conjuncture" in Llewellyn Gross ed., *Symposium on Sociological Theory*, Harper & Row, New York, pp. 152-3.

Tuma, Nancy Brandon, Hannan, Michael, T., & Groeneveld, Lyle, P., 1979: "Dynamic Analysis of Event Histories" in *The American Journal of Sociology*, Vol. 84, No. 4., January, 1979, pp. 820f; 821-2; 847-851.

Vidich, Arthur, J., 1955: "Participant Observation and the Collection and Interpretation of Data" in *The American Journal of Sociology*, Vol. 60, January, 1955, p. 36.

Wall, W.D. & Williams, H.L., 1970: *Longitudinal Studies and the Social Sciences*, Heinemann for The Social Science Research Council, London, pp. 8; 15-17.

Whittle, P., 1968: "Time Series: II, Advanced Problems", in *International Encyclopedia of the Social Sciences*, ed., David L. Sills, The MacMillan Co., & The Free Press, U.S.A., pp. 59, 60.

Zerubavel, Eviatar, 1979: *Patterns of Time in Hospital Life: A Sociological Perspective*, The University of Chicago Press, Chicago.

Chapter 5

Reality in Social Research

In this chapter we shall be positing what are, perhaps, *the* basic questions in social research. Does the researcher apprehend and *re*construct a given social reality, thereby attempting to explain that reality?; or, alternatively, does he simply create his own reality, which constitutes one of many possible interpretations; that is to say, one of many possible constructions of reality, including the mundane, the scientific, or the mystical? If we were to accept the latter proposition, namely that social research does not recreate, it does not reconstruct, reality, but it creates its *own* reality, then this scientifically created reality can never meet the day-to-day realities that it is ostensibly interested in describing and explaining; if we assume that such a day-to-day reality exists independently. Even if we were to allay such fears by the contention that this proposition does not deny that the social researcher's reality has a certain affinity, even a close one, to some assumed everyday life, it would still put him into an inexorable position. He would be akin to someone travelling alongside the tracks but never managing to get onto them. A situation of this kind would seem pretty hopeless for social research.

We have posited here a vexing problem: one that has given rise to a good deal of controversy. We cannot claim that the discourse we shall develop in the coming pages will lift the entire issue out of the realm of uncertainty in which it is still entrapped. However, after examining a number of alternative views an attempt will be made to present a plausible idea concerning sociological research and its relation to reality. To begin with, we hold that such research is not a *pure re*construction of reality; but that it is neither simply the creation of a separate and essentially different reality from the everyday world. It is, rather, a specific construction of reality resulting from the interaction between disciplined scientific thinking and everyday life, the intervention of the former having the purpose of exploring and explaining everyday reality. But in so doing it appears that science, through its concepts, models and theories, construes the facts it studies in particular ways. If so, taking the sociological field as an example, how are we to select out from the different social scientific constructions of reality that which is true? Or put another way, can science get at what is true reality, and how do we define 'reality' and 'true reality'? Can we assume, for instance, that true reality is that ultimate reality which may be ascertained objectively?

The sociology of knowledge perspective

An excursion into the sociology of knowledge will be necessary here, and we might as well begin with a note regarding the distinction that Karl Mannheim, Max Scheler and Kurt Wolff make between 'scientific truth' and 'existential truth'. The former they regard as stipulative and propositional, which is a pragmatic truth required methodologically for the solution of scientific problems. Truth in this sense is a 'means' criterion. On the other hand 'existential truth' concerns volitional and metaphysical problems and it represents an 'end' criterion; it is essentially experiential and intrasubjective. This dichotomization leads Kurt Wolff to the statement that "science makes no claims about the nature of ultimate reality; it is not concerned with this reality" (Wolff, 1970). And in an identical vein, Scheler states; "Metaphysics .... attains its goal, which is knowledge of essences, by essence-visioning reason, not by observation and inference; while science does by observation, experiment, induction, deduction" (Scheler, 1970). Karl Mannheim similarly asserts that the scientific attitude dispenses with the ontological quest.

And although Mannheim develops the idea of the underlying social or existential determination of thought, he exempts the exact or natural sciences, including mathematics, from the verdict of existential determination; this in contrast to "historical, political and social science thinking as well as the thought of everyday life" (Merton, 1957, Stark 1958, Mannheim 1936). Yet, presumably, even in the natural sciences a scientist may deviate from the 'pure type' insofar as there may be "metaphysical premises which have their influence on the selection and formulation of problems, on the interpretation of findings, etc." (Wolff, 1970). Scheler stresses more inclusively the social character of all knowledge but not in terms of its content or objective validity, but in terms of the "selection of its objects according to the *ruling social interest* perspective" and in that "the mental acts by which knowledge is won are always and *necessarily sociologically* co-conditioned, i.e., by the structure of society" (Scheler, 1970). He rejects, however, sociologism by distinguishing between the *content of judgments* determined variously by his metaphysical 'timeless essences', and the *act* of judgments determined by historical and social reality (Scheler, 1970, Merton, 1957).

The dilemmas reflected in these views concern the extent to which a sociologistic doctrine, claiming the social determination of scientific knowledge, can be avoided, and by implication they lead us to question the basis of scientific objectivity. It would appear that the limitations of science in this respect must be recognized, at the very least by accepting the Kantian notion, repeated by Max Weber and others (Eldridge, 1971), that science cannot be presuppositionless. But we shall develop the argument below: (a) that if so, lack of complete objectivity is the fate not only of social science but of all science; and (b) that, nevertheless, the limitations regarding scientific objectivity need not destroy its rational basis.

Thus, we can demonstrate that the social sciences are not alone in the situation of not being able to attain complete objectivity. Contrary not only to popular belief, but also to the erroneous assumptions in some academic circles, we shall argue that the physical sciences are not any less beset by problems of non-objectivity than the social sciences. What follows is neither apologetics on behalf of sociology, nor is it to be taken as brave words of encouragement. It is an attempt to put the record straight. Physics can serve for us as an excellent example, since it is one of the disciplines which, especially in its modern development, has been much influenced by philosophical and social scientific conceptions of reality, and the concepts of which, in turn, have strongly influenced all the other sciences.[1] As Max Jammer points out, Niels Bohr, for instance, was greatly influenced by William James' notions of psychology, especially by his ideas of the nature of the 'stream of thought'. Bohr's writings themselves indicate how his theories, and particularly the epistemological implications, were inspired by James' ideas. The corss-fertilization between physics and philosophy, and to a lesser extent the mutual influences of the social and natural sciences, have been highlighted and discussed by many writers (Jammer 1966; Capek 1961; Colodny 1962).[2]

In attempting to understand what the nature of physical reality is, physicists have constructed models which might have brought them to "the edge of objectivity" (Gillispie, 1960), but which, time and again, proved to be inadequate or incomplete (Capek, 1961). This is basically due to the abstruse nature of physical reality. During the last century thinkers like Cournot thought that "natural laws as construed by science are only approximations since physical reality itself is subject to contingency; an event being contingent if its opposite involves no contradiction"; whilst C.S. Peirce simply declared that "nature is not regular" and that "chance is a factor in the universe", (Jammer 1966). Deterministic physical science was on the way out by the earlier part of this century.

The development of quantum physics culminated in a science based on the indeterminacy principle, on uncertainty relations and on probability laws, as reflected in the ideas of Niels Bohr and Werner Heisenberg. Recourse to statistical probability was forced on this science by the discovery of the discontinuous nature of atomic radiation; the pulse-like process in the

[1] F. Znaniecki (1952) points to a similarity between physics and sociology, in that physics is the basic natural science and sociology the basic cultural science.

[2] It may be noted, however, that the drawback in the discussions between physicists and philosophers has also been mentioned (Feyerabend, 1962).

emissions of energy. A radiating atom does not deliver up its energy continuously, but discretely in bundles; furthermore, prediction becomes uncertain since "quantum theory can give an indication of the probability that the alpha particle will leave the nucleus in unit time, but it cannot predict at what precise point the emission will occur". To add to difficulties, it was discovered that it is impossible to describe simultaneously both the *position* and *velocity* of an atomic particle with any prescribed degree of accuracy. One can *either* measure the position very accurately when the action of the instrument used for the observation obscures one's knowledge of the velocity, *or* one can make an accurate measurement of the velocity and forego knowledge of the position (Heisenberg, 1930; Heisenberg, 1958).

Henry Margenau believes that this dilemma exists in the social sciences too (Margenau, 1950). Indeed, analogous situations may be found in abundance in empiricist social research. Thus, certain types of participation-observation studies provide us with reliable information about social action, but they give us limited views of the whole society; their findings are not representative and do not enable us to make reliable inferences about social structures (Douglas, 1973). Survey techniques, on the other hand, provide a better overall picture, but are unlikely to be able to delve into the meaningful interpretations of social relations by those engaged in informal practices. To superimpose the two techniques on each other might conflate evidence from different social situations, which is likely to blur rather than clarify the nature of social relations. Yet the use of only one of the techniques will provide partial information.

In a similar vein we have referred in Chapter 2 to the difficulties in achieving the same level of measurability of different indicators relating to a particular variable, and of establishing the exact relationships between the various indicators. Despite the basic differences in the nature of the subject matter of natural and social sciences, heuristically they have much in common. Of late, the natural sciences have been borrowing models from the social sciences: in biology the genetic code is described in terms of linguistics and biologists also speak of cell sociology in order to explain communication between cells as a phenomenon akin to social process (Levi-Strauss 1976 and 1979). And, as we have seen in Chapter 3, natural science models are sometimes used in the social sciences. Given the signs of congruence between the sciences in many respects, not least in the matter of non-objectivity discussed below, Carl Hempel's stand concerning "the methodological unity of all empirical science" is understandable (Hempel, 1962).

In order to stress further the elements of non-objectivity and uncertainty in science, let us buttress our brief account of developments in modern physics, presented above, in a layman's language, by a summary of Werner Heisenberg's own statements. Heisenberg holds that so far as the physical world is concerned it is not possible to ask whether the elementary particles of nature exist in space and time *objectively*. According to his philosophical exposition of developments in atomic physics the situation is that in order to form a picture of the nature of the elementary particles, one cannot ignore the physical processes through which knowledge about them is obtained, since processes of observation of the smallest building particles of matter produce large disturbances. Then, quantum theory does not deal with the elementary particles themselves, but with a knowledge of them. Hence "the only processes we can refer to as taking place are those which represent the interplay of particles with some other physical system, e.g., a measuring instrument". Therefore, "the atomic physicist has had to resign himself to the fact that his science is but a link in the infinite chain of man's argument with nature and that it cannot speak of nature in itself". The scientist does not produce a picture of nature but a picture of his relationship with nature. It is unsuitable, as a starting point in modern science, to speak of objective processes in space and time or to regard the scientist as an objective observer confronting nature. Thus, "the scientific method of analysing, explaining and clarifying has become conscious of its limitations, which arise out of the fact that by its intervention science alters and refashions the object of investigation. The scientific worldview has ceased to be a scientific view in the true sense of the word ...." (Heisenberg, 1930; Heisenberg, 1958; Heisenberg, 1975).

The context of these incisive statements by Werner Heisenberg is a more inclusive analysis which eschews strict determinism in science and in which the elements of non-objectivity and uncertainty in scientific work are clearly recognized. Karl Popper has argued against Heisenberg's and Bohr's subjectivist interpretation of quantum mechanics, on the ground that the latter is a statistical theory, "*because* the problems it tries to solve-spectral intensities, for example, are *statistical problems*" (our italics). Popper then sums up his position in an inconsistent manner: on the one hand he recognizes "the irreducibility of statistical theories to deterministic theories"; yet he maintains that the two kinds of theory are not incompatible and that there is no reason whatsoever to doubt "the realistic and objectivist character of all physics" (Popper 1972). But taking into account his own statements, it is surely logical to say that since statistical theories are part of physics and since such theories cannot be reduced to deterministic theories, there remains an area of indeterminacy in this science, precisely as Heisenberg argues. These characteristics of indeterminacy appertain with equal, if not more, force to the social sciences; something that should appear natural if the basic unity of science is recognized. We are not simply trying to assert by declaration, so to speak, the applicability of Heisenberg's views to the social sciences. We have argued for and demonstrated such views all along: when we presented variables as mediating constructs between the everyday phenomena of the social world and scientific notions; when we pointed to the unavoidability of fractional measurement; when we discussed the problematic nature of rules of correspondence between the conceptual and the empirical, (see also below); when we highlighted the complexity of human actions and our inability to take a deterministic line in causal analysis; and when we stressed the significant role that the time factor plays in sociological research and explanation.

The relationship between science and reality

Following Heisenberg's notions enunciated above, which we maintain hold good both for the physical and social world, we must ask additional fundamental questions: namely, in what sense, if at all, is science concerned with ultimate reality and, more particularly, what is the relationship between reality and knowledge. But first, what is 'ultimate reality'; how can we define it? We could accept the idea of ultimate reality as being some objective reality 'out there', so to speak, whether physical or social, which holds a non-context dependent truth. We could present a variety of views, from Descartes to Kierkegaard: from the Cartesian notion of a reality which arises from one *indubitandum*, our thought processes leading us from *res cogitans* to *res extensa*, that is from thinking to existence; to Kierkegaard's insistence on the antithesis between thought and reality, since "thought could never attain reality, for as soon as it thought to have done so it falsified reality by having changed it into imagined reality" (Jammer, 1966; Castell, 1976). We may also mention Scheler's view of the existence of an 'underlying' reality other than thought to which thought is always related (Berger & Luckman, 1966; Scheler, 1970), or Levi-Strauss' view that the underlying structure that pervades culture and the human mind is reflected in language, which is co-extensive with social reality (Levi-Strauss, 1976). Yet others label 'second-order realities' those realities which are socio-culturally constructed and are imparted with various measures of precariousness, and regard them as different from the more solidly based first-order phenomenal features of everyday social life, the former "*comprehending from an analytic distance* the nature of 'everyday matter-of-factness'" (Robertson, 1972). One can also contrast, more simply, the 'fictive', artificial, man-made world as against the natural world (Simon 1969; Holzner, 1968). Karl Popper highlights the difference between man-made theories and the *real facts* (our italics) which, according to him, are "with very few exceptions, emphatically not man-made".

Apart from the difficulty of accepting from a sociological standpoint, the idea of real facts not being man-made, if we consider for the moment meaningful social interaction and its consequences, in his pluralist framework of the three realities, quite surprisingly, the very man-made theories he highlights, emerge as constituting the autonomous and objective *third* world. For Popper the physical world, that which kicks back at you, is 'world one'; 'world two' consists of psychological processes of feelings and subjective experiences; whilst 'world three' contains the products of the human mind, theories and problems, science and knowledge. All three worlds are *real* and, although distinct, they influence each other, they interact. (Popper, 1972) Our very brief review of the variety of views concerning reality, that is, its true or ultimate nature, reveals an area of interest to those seeking to understand the link between scientific research and reality. One thing is clear: whichever of the above views we adopt, science must be seen as an integral part of and intimately concerned with ultimate reality; quite contrary to Kurt Wolff's view cited earlier. The position of science is such both at the empirical level, through its obervational and experimental interventions, and at the cognitive level, through its conceptualizations and theorizing.

We can posit now the following set of assumptions about our relationship, as scientists, with ultimate reality. First, we may assume that we shall never know fully what ultimate reality is; there is no way of uncovering the final non-context dependent truth. We are, to a marked degree, context bound and reality is ultimately too complex for our abilities to fathom it fully. A corollary of this would then be that as we attempt to uncover bit by bit, one aspect or another of this ultimate reality, we give various interpretations to it; hence knowledge of it is hermeneutic; it represents stages, degrees, aspects, vantage points, reflected in scientific findings and subjective evaluations and meanings, of that ultimately unknowable reality, or at least a reality unknowable in full. Even so, implicit in this set of assumptions is the view that there does exist out there, some objective reality which we seek to explain and understand. Ultimate reality in this sense *is* the objective reality, insofar as it is "independent of our own volition" as Berger and Luckman say, in that we cannot "wish it away". All those searching for an understanding of the nature of physical reality were similarly inspired by a feeling of the objectivity of physical reality "whose structure exists independently of our preferences and idiosyncracies" (Capek, 1961). Even the transcendental, as opposed to the physical and usually held to be irreducible to other existences and realities, once conceived, it represents a reality, since the perceiver is bound by its very conception, which is its existence.

Werner Heisenberg also recognizes that in view of the imprecision of all perception "it may be suggested that behind the statistical universe of perception there lies hidden a 'real world' ". But he dismisses such speculations as useless and meaningless (Jammer, 1966). Furthermore, if we follow his notions, we have to stress the very close link, tending to a point of merger, between reality and knowledge. Berger and Luckman are careful not to allow themselves to be bogged down in a philosophical stance that would make reality appear to be independent and separate from knowledge; i.e., the latter simply representing attempts to verify the distinct 'real' entity. They do deal with this distinction but at the same time they stress the relationship between the two.

In fact they justify the preoccupation with the 'sociology of knowledge' by the need to examine "the processes by which *any* body of 'knowledge' comes to be socially established as 'reality' " (Berger & Luckman, 1971). In this sense, we arrive at the idea of multiple realities, since reality is always socially, that means variously, constructed. Coming thus far with Berger and Luckman we could easily have dovetailed their ideas to the notion that knowledge and reality intertwine and become one where science is concerned: the atomic reactor is both knowledge and reality and so is a set of birth rate figures of a country.

The interrelationship may be further stressed by the most self-evident proposition that we cannot know, judge or apprehend without there being something to know, to judge, to apprehend (Meinong, 1902). However, there are certain aspects in Berger and Luckman's 'sociology of knowledge', which, despite its phenomenological stance, with its emphasis on the subjective process and meanings creating the *inter*-subjective common-sense world, echo, albeit in a restrained tone, the notion of an *objective reality*. They use the terms "compelling facticity of everyday life" and "paramount reality" to convey the idea of a more enduring and encompassing reality, which envelops and serves as a point of departure and return for the specialized realities of science, religious mysticism, dreams, or aesthetic experience. Then, in Schutz's style, they describe the leaps of consciousness which are necessary for commuting between these "finite provinces" and the paramount reality (Schutz, 1972; Berger and Luckman, 1971).

For the purpose of rational understanding the value of this distinction is doubtful. One can throw much more light on the relationship between the many realities by adopting some of the concepts developed by Mary Douglas in her anthropological studies. Thus we could argue that there are various universes of knowledge with their favoured patternings of reality and with their implicit categories, one such universe being the scientific one. What Douglas says is that both explicit rules in a society and the implicit meanings stem from the experiences and interests of that society which are channelled through "socially significant categories". This Durkheimean notion of the social determination of culture and knowledge is extended by Douglas to include the sphere of science. In this connection she quotes Wittgenstein whose lack of faith in objective scientific truth lead him to assert "that even the truths of mathematics are established by social process and protected by convention" (Douglas, 1975). Her argument that all categories and all meanings are social, would appear to be no different from Berger and Luckman's insistence that reality is socially constructed. Where the two approaches do differ, however, is that whereas Berger and Luckman stress the cleavages between the various realities, Douglas insists on the existence of an overall unity stemming from basic social boundaries, categories and meanings (Berger & Luckman, 1971, Douglas, 1975). There is a similarity between her ideas and those expressed by Speier and Znaniecki concerning the influence of socio-cultural patterns on systems of knowledge and on the scientific methods built by scientists (Curtis and Petras, 1970).

The question now arises in what way science, in our case social science, is anchored in social experiences and interests. How are its paradigms, categories, and concepts, in fact the whole gamut of its methodology, related to social reality? An answer is apparently provided by Jurgen Habermas who sees the link between scientific inquiry and reality in terms of *knowledge constitutive interests*. Namely, the problems of the preservation of life have been solved by the cultural form of existence. This is so, since work and interaction include processes of learning and mutual understanding, and these processes "have to be maintained in the form of methodical inquiry". This does not mean, Habermas tells us, that knowledge is "a mere instrument of an organism's adaptation to a changing environment", but neither is it, according to him, "the act of a pure rational being removed from the context of life in contemplation" (Habermas, 1972). Yet Habermas retains a comprehensive rationality which he links to self-reflection. For what he claims happens in the process of inquiry is that we are "moving the whole time through the spectrum of reasonable discussion". It is his wish to widen the horizons of knowledge by arguing against the positivistic division of rationalism, i.e., reserving the rational approach to the purely technical and practical aspects of inquiry and thus leaving the realm of ends to pure decisionism. For an explication of his approach he proposes a trichotomy of human interest in which knowledge is anchored, reminiscent of a Comtean ascending order of human knowledge. The three types of interest are: the technical interest of the empirical analytical sciences; the pratical interest of the historical hermeneutic sciences; and the emancipatory interest of critical theory. It would appear then that the purely technical interest of empirical methodology is restrictive and is abandoned where the social life-world is submitted to the natural hermeneutic explication of meaning. His final aim is to unify theory and practice by means of the dialectical concept of totality. In this sense, the dualism of cognition and evaluation of facts and decisions, must be transcended, since society is to be conceptualized as a "historically emergent totality" (Habermas, 1974, Albert, 1974, Habermas, 1972). We shall present below some criticisms of Habermas' ideas in connection with the *positivismusstreit* in which he engages, leaving aside the nagging nature of his convoluted statements, and concentrating on the boomerang effect that his ideas have on his own scheme and philosophy. But Habermas' insistence on a 'total view' does direct our attention to the mutual interaction between social scientific research and ongoing social life. It is, thus, clear that the research process is not pure, having been influenced by the preconditions of understanding and knowledge and that scientific research itself affects reality by means of its results and applications.

Ethnomethodological and positivistic approaches to social reality

Scientific research could, therefore, be regarded as a process that creates reality and even more obviously as a process that changes the meanings of reality. Science, that is to say the knowledge it provides, helps us shape the meanings that nature and its objects and social life have for us. In this sense reality is what we make of it, whether as non-scientists or as scientists. The ethonomethodologist's standpoint that such meanings created by empiricist-positivist science are reifications of reality and do not represent the real meanings found in everday life is an argument which leads to an untenable position for sociology. According to this argument "All social meaning is negotiated, and there is no stable connection between one situation and another similar one other than the conventional label we apply to classify them" (McSweeney, 1973). And since science classifies, categorizes, unifies and explains by reducing entities to their common properties and structures, it loses in the process the real meanings which are negotiated, ad hoc, by members of society and which are known only to them. So goes the ethnomethodological thesis. As Bill McSweeney so forcefully argues, this leads to sheer absurdity. Some useful insights notwithstanding, phenomenologists have unwittingly produced ethnomethodologist progenies who in their furious attacks on scholastic realism and functional prositivism have reduced sociology to the sterile activity of decoding and encoding context-specific meanings. Such activity is not intended to add up to anything beyond endless, and senseless, description; thus leading to an overall state of meaninglessness in sociology (McSweeney, 1973).

We shall pursue, for a little while longer, the critique of ethnomethodology and thus arrive at what appears to us to be the link between meaning, reality, and social science. If ethnomethodologists feel that sociology is "a cultural device whose theories and methods endistance us from the everyday, the apparently trivial, the shifting, fine-grained,tricky existence of actual people experiencing and creating reality in the minute contexts of personal fleeting intercourse", then Donald MacRae is right in pointing out that it is precisely at this level that "cohesion (and disjunction) are most fundamentally found" (Mac Rae, 1972). It seems reasonable to hold that social cohesion as well as disjunction are the results of meaningful interaction and that the cohesive, the disjunctive, and the conflictual situations are all meaningful in themselves. Such a view does not necessarily endorse the existence of "a stable meaning overarching a variety of different contexts". Yet it is inconceivable that there should be no stability whatever, no objective element at all, in any situation. As McSweeney asks, what possible negotiation about meaning can take place without some basis of prior agreement? There must be common ground for either agreement or dispute. It is such a common ground that is always constructed among human beings and this is the social reality that the sociologist seeks to understand.

McSweeney points out some of the diversity within ethnomethodology, yet he deals with its main thrust, and he is most convincing when he focuses on the naivete and the fallacy of attempting "to negotiate meaning across a vacuum"; of attempting to obtain 'raw' data from scratch, so to speak. Such attempts arise from the mistaken notion that it is possible to uncover the 'pure' facts underlying the reifications and constructions at the common-sense level as well as at the sociological level, namely, the underlying layers of socio-cultural reality. There "can be no knowledge of particular things without abstraction of properties common to things already known, so there can be no grasp of particular situations without abstracting from the particular or unique character of the event meaning-structures already culturally given". Meaning, therefore, is "an interpretation within agreed boundaries of relevance"; it does not preclude *given* or *ready-made* accounts of reality (McSweeney, 1973; Gellner, 1973). In this sense, meaning is not "an independent, free floating entity, which we try to capture more or less completely" (Quine, 1976). It is partly given and partly newly constructed and it is through its social scientific interpretation that reality can be partially understood. It is, therefore, the task of the sociologist, as social scientist, to pursue abstraction, generalization and theorizing so as to arrive at "overaching decontextualized meaning structures". Yet it is also legitimate for the sociologist "to assert the context-specificity of meaning in certain areas of social action", so long as sociological analysis is not reduced to the triviality of describing unique indexicality unrelated to any overall social structure. (McSweeney, 1973).

Now let us sum up this critical note of ethnomethodology and consider its implications. The crux of the matter in ethnomethodology is that since context-dependent indexicality, that is: the momentary, unpredictable, subjective, and forever negotiable expressions of individuals, represent the *real* concrete situations, these may be regarded as being in a completely fluid state, hence the decoded de-contextualized objective expressions which are substituted for them are misleading or false, since they do not represent reality, they obfuscate it. That is to say, the decoded expressions are 'remedied' representations, glossing over contextual details. They are, therefore, unacceptable. Since the context is never closed, being always open to revision and redefinition, and since meanings are always negotiated afreash, ethomethodologist sociologists insist on entering, as virtual participants, the context in which the action and its meaning is embedded. As McCarthy, following Habermas, puts it: "The methodological implications of this thesis are far-reaching: 'If a context of action can be opened up only from within, that is through (at least virtual) participation in a communication that is set within, and at the same time further develops this context, the sociologist must in principle assume the same position as the lay member. He has *no privileged access to the object domain*' " (our italics). Since we may insist that no methodological system is privileged, it is legitimate to claim that ethomethodologists cannot dissolve their own situation, hence they cannot reach the unreachable goal they set themselves (McCarthy, 1978, Capek, 1961).

Ethonomethodology in its quest for apodictic knowledge places itself in an ineluctable position. This is not due to its metaphysical approach of claiming to probe underlying reality as opposed to superimposed appearances, something that science also claims to do. Although we need not enter a demarcationist dispute between science and metaphysics, it is still true to say that the latter's greater emphasis on understanding ultimate origins and absolute truth, has apparently influenced the ethnomethodologist in his attempt to break through to the real essence or the pure meanings and sources of social life. In this endeavour ethnomethodology finds itself in a trap: it wishes at one and the same time to uphold the critical attitude of metaphysical philosophy and yet to claim a privileged frame of reference. But there are no grounds on which it can arrogate to itself such a methodological position. Since the philosophical grounds on which ethnomethodology is built are untenable, and since research activity within its framework is likely to boil down to sterile and uncheckable descriptiveness, it is advisable to discard it, for it is not a viable sociological methodology.

What is the alternative we are left with? Is the methodological framework of sociology simply wedded to the empiricist scientific quest for rational explanation, and does this mean that we must adopt, what Imre Lakatos calls, 'instant rationality': that is, the notion that crucial experiments and research findings will instantly give us scientific truth (Lakatos, 1978; Hacking, 1979). Is empiricist science claiming a privileged position, in the manner of Popper, by defining scientific truth as "correspondence to the facts" (Popper, 1972)? We may argue that such absolutist claims are not tenable and that instant rationality must be cast out and be replaced with the more moderate claim of 'having good reasons for' accepting a scientific theory; this being equivalent to a 'high degree of probability' (Carnap, 1950; Schilpp, 1963). More moderate claims are forced upon us since scientific theory cannot be absolutely verified. We are never able to scrutinize all possible instances nor do we have the means of "completely reducing theoretical predicates to empirical or 'observational' predicates" (Tondl 1973). The problem of what is 'corroboration', 'falsification' or 'refutation', will forever remain with us. We cannot know, therefore, how fully, how accurately, or how truthfully science represents reality, if we were to adopt a stance of *representational realism* for science (Blackmore, 1979).

Again, we shall always have the problem of determining what are positivistic and deterministic excesses. Thus, do we have to reject the 'myth of the given' and must we unequivocally declare ourselves against any objectivist illusions? Not if we agree with Karl Popper's philosophy. Popper upholds the relativistic thesis of facts being constituted within frames of reference or horizons of expectations but argues, nonetheless, for the retention of an objectivist notion of facts and knowledge. As we have mentioned above, within his pluralist framework of the three worlds of reality, the objective scientific reality of the 'third world' emerges and is constituted through the intersubjectivity of the scientific method and its public character. But despite the claim by Popper and Lakatos of the existence of an autonomous, independent and objective area of knowledge, that is scientific knowledge and its growth, and despite Popper's protestation that the principle of "indeterminism is not enough" if we wish to allow for man's freedom and creativity, Popper does partly capitulate to the Heisenberg principle. Thus, he acknowledges that chance events in quantum physics are of a much more radical nature – they are absolute; they differ from the chance events due to the independence of two causal chains which happen to meet, interfere, or combine at some point in time and space. However, Popper *can* retain an objectivist stance since he may claim that there is nothing subjective about absolute chance. As for man's freedon and creativity, he maintains that although the human being is part of nature, he can transcend himself and nature by creating the 'third world' of knowledge, which is a product of scientific activity. That is to say, although knowledge contains products of the human mind it becomes independent of it and emerges as an objective reality. Yet Popper admits that the universe is a complex mixture: "partly causal, partly probabilistic and partly open" (Popper, 1972; Popper, 1973; Lakatos, 1970; Hacking, 1979).

Rationality, objectivity and testability

This part openness, this trancending capacity and the ability to engage in rational discussion are of crucial importance to science in general and to social research in particular. Our ability to transcend everyday reality through the research process involves a kind of 'pulling ourselves up by our own bootstraps' exercise. This is possible and necessary, since as David Martin puts it: "Man is neither so context-bound as an animal nor so context-free as an angel" (Martin, 1973). We are not fully objective as human beings, the physical and human worlds evince some regularities but are also chance-ridden, our context-boundedness puts serious limitations on us, yet we can probe the world around us and attempt to understand it better. Above all, we need not impugn our ability to engage in logical or rational action and thought, although we are far from being purely rational beings. Some argue that we possess 'natural rationality', that we have a proclivity for natural reasoning, which is based on inductive propensities. Another way of seeing man's rational powers is by pointing out that although he always applies schemata to the world, his mental activity also contains a corrective feedback, making him into "a tester and processor of hypotheses even when he is unreflective about [it]" (Barnes, 1976). In science we operate on such rational principles consciously and we invoke rationality in the sense of "the critical discussion of alternative techniques of enquiry, competing theories, various definitions of basic predicates, in short, of different frames of reference and their elements". Those who are not convinced by the psychological concept of 'natural rationality' need not conclude, as Popper does, that the basis for accepting such critical rationality is necessarily irrational. His own methodological writings "are replete with good reasons for adopting it, reasons based on understanding of the process of enquiry as a whole in relation to basic structures of human life" (McCarthy, 1978). Perhaps the very fact that the social scientist is prepared to question the rational foundations of science is the best proof of his ability to think rationally. The anti-rationalist stand is not as formidable as Roland Robertson imagines (Robertson, 1972). Attempts to deny the validity of analysis and interpretation can be met with satisfactory counter-arguments. One such argument is the idea of the growth of knowledge, which in Lakatos' scheme is the demarcation between 'rational activity' and 'irrationalism' (Hacking, 1979; Lakatos, 1978).

This demarcationist criterion is more difficult to apply to the science of society. The reason is that some sociological knowledge is characteristically hermeneutic, and therefore, its growth and additiveness cannot be clearly demonstrated (Hacking, 1979, Habermas, 1972). What is more, there is an element of incommensurability which poses a serious problematic. Various sociological research results and different theoretical propositions often explain reality in different ways. Two distinct sociological analyses, even with identical research data and the same analytic technique, may take different lines and arrive at what appear to be contradictory conclusions. How does this arise? We may exemplify here with two differently constructed path models using the same correlation data. The two models can vary in the causal relationships they show, since the models used are subject to the theoretical definition of the situation, which will decide *a priori*, rather than on clearly substantive grounds, what precedes what. Such a procedure is likely to occur because in sociological research the independent variables are, as a rule, not controlled or manipulated; there is often no clear time sequence showing causality; and many simplifying assumptions are necessary (Duncan, 1975, Asher, 1976, see also Chapter 3). Hence it can be asserted that all sociological explanations are only partial since no one can possibly obtain a full account of reality. There may, therefore, be no real or necessary inconsistency, but rather a situation of incommensurability. Consequently, since no one has the last word, does this mean that effectively, and especially in relation to social science, anarchism rules in scientific research? (Feyerabend, 1975). Or looked at another way, is it the complex reality beset by compound, multiple and parallel causations that gives the rise to the difficulties? And are these to be overcome eventually by the research programme and its "heuristic", defined by Lakatos as "a powerful problem-solving machinery" capable of digesting anomalies (Lakatos, 1978)? Alternatively, can the apparent inconsistencies be explained by Habermas' view that scientism is an ideological distortion since its claim of the unity of scientific method "ignores the constitutive interests of the enquiring subject and the social context within which these arise"? (Overend, 1978; McCarthy, 1978).

Habermas' rejection of the 'objectivist illusion' and his strictures on positivism are not novel. They could have been added to the many warnings against a false sense of scientific security. However, his critique is linked to a questionable alternative. He appears to clamour for, in fact claim, an exclusive access to the foundations and formation of knowledge. His basic aims are: (a) to open up the concrete historical totality which is to be understood as an objective meaning context; (b) to understand the meaning context with the aid of a theory of communicative development and competence. If the grounds of his critique of positivism are that the latter does not *reach behind* "the profound historicity of human thought and action", (McCarthy, 1978) then Habermas lays himself open to the boomerang effect of his own critique. Let us look a little more closely at the issues involved so as to understand why this is so. Habermas unfurls the elements of his methodical approach by declaring a need for a dialectical 'thinking through', guaranteeing the appropriateness of categories as one proceeds, arriving thus at reconstructions, beginning from everyday language and the stock of everyday knowledge and culminating in a critical theory which takes into account its own predilections and is thus emancipatory (Habermas, 1972, Habermas, 1974, Giddens, 1974). He attempts in this way to leave behind layers of subjectivity, in order to arrive at objective meaning. But can he achieve such a breakthrough? Is he not unavoidably caught between an 'infinite regress', not unlike the ethnomethodologist, and an *a prioriness*, not unlike the positivist-empiricist? Can he avoid choosing one or the other? The answer is 'no'. For if he is to gain access to a particular life-world to achieve his systematic reconstruction he must do this 'performatively', that is to say as a hermeneutic interpreter, which means participating in a process where as an interpreter he comes into contact with the 'object' being reconstructed (McCarthy, 1978) This would plunge him into a hermeneutic whirlpool, for his own reconstruction would have to be subjected to another reconstruction of what went on in his earlier reconstruction, and so on, leading to the 'infinite regress' (Giddens, 1976; Tilley, 1980).

Alternatively, he must decide in an arbitrary *a priori* manner the spatial boundaries and erstwhile beginnings of the situation, act, knowledge, etc., being reconstructed. The possibility of free floating values and the idea of moving boundaries and changeable time relations, as derived from relativity theory, have not affected the empiricist's seeking of closure. This is necessary in order to idenify one's objects of enquiry (Lachenmeyer, 1973) a necessity Habermas may not be able to avoid in his reconstructions. Hence McCarthy fears that Habermas has, in the final analysis, not been able to solve these methodological problems. In fact, Habermas' approaches provide the most recent 'objectivist illusion', one that could be unmasked by exhibiting the context-boundedness of basic categories and assumptions (McCarthy, 1978). To sum up, Habermas remains wide open to the attack that must be faced by anyone with a priviliged access claim. His criticism of positivism's operationalist confidence in science's ability to gain access to a factual reality, despite the inherent uncertainties of such scientistic claims (see below), reminds us of the selective attacks launched oftentimes by writers whose own theories are not immune to such attacks (Gellner, 1974). But even worse, as Hans Albert says, Habermas nowhere provides a demonstration of the methodical advantages of his ideas. Verbal exhortation of 'totality', 'transcendence' and 'breakthroughs to objective meaning' are insufficient (Albert, 1974).

In contrast, the Popperian scheme, despite its own problematics provides a clear idea of how objective scientific knowledge of reality is to be gained. Anthony Giddens makes the point that there is a terminological confusion around the positivism dispute between Popper and Habermas. The former claims not to be a positivist since he does not hold that science is simply generalization based on observation, and since he has always stressed that all observation is theory-impregnated, so that one can only check and refute but not prove theories. The latter on the other hand recognizes that Popper rejects 'empiricism', which is actually close to what the former means when he uses the term 'positivism', but insists that Popper retains a positivistic stand since he "assumes the epistemological independence of facts from theories", and since he approves the testing of theories by means of facts, despite the inherent 'uncertainty' of scientific information (Giddens, 1974). Since, in our view, the term positivist ought not in itself be deprecatory, what matters is the usefulness of a scheme, and this is where Popper has the upper hand. The two basic elements in his methodology are bold conjectures and testing (Popper, 1974; Popper, 1972). Sociological work, far too often, fulfils the first requirement and not the second. If the element of testing is to be carried out more adequately, it is a *sine qua non* that the applied sociology angle will play a vital role.

Why is testability essential and how is sociology to achieve a greater measure of testability? The importance of testability is clearly set out by Ted Benton: "Contemporary scientific knowledge is represented as the sum of all those propositions which have survived generations of repeated confrontations with reality. Empirical testability is the mark of the openness of claims about the world to revision under the impact of the world itself" (Benton, 1977). But this empiricist conception of scientific knowledge is fraught with serious difficulties. Radical empiricism holds that we must clearly distinguish between observation statements and theory statements. The two become linked for purposes of scientific elucidation by means of mixed statements, usually called 'bridge principles' or 'correspondence rules'. These latter are a composite of theoretical and empirical terms and they provide partial interpretations of the theoretical element in science. As such they are instrumental in rendering the theoretical element scientifically legitimate, since they reduce the proposed scientific theories to observable consequences. Thus, theories concerning the behaviour of gases and the actual observation of such behaviour are linked by *bridge principles* such as kinetic energy, mean speed, concentration, etc., (Benton, 1977). An example in sociology would be the linking of a theory of social domination and the observation of the social immobility of the subordinate race, by means of bridge concepts such as prejudiced attitudes or class interests.

It is clear that it is this procedure, as conceived by empiricists, which makes testing feasible. Yet many questions have to be faced. Fundamentally, it is not at all clear where we draw the line between the 'observable' and the 'unobservable'. Are things like aggregate supply or national income mere econometric calculi and, therefore, not directly observable as empirical phenomena? (Rosenberg, 1974). Or, are solidary relations really observable? How do we distinguish in such cases between the experiential and the theoretical? The lack of clear criteria for such distinction appears to be due to the impossibility of defining in purely experiential terms the statements of empirical science. This must mean that testing itself will never be complete. Thus, Rudolf Carnap in his later more liberal positivist position says: "The interpretation of theoretical terms is always incomplete, and the theoretical sentences are in general not translatable into observation language" (Carnap, 1956). Others, like Hempel and Machlup, reflect the same softening of the empiricist stand by emphasizing that the greatest advances in scientific systematization have come about through hypothetical or theoretical thinking rather than by direct reference to observables, and that basic postulates are not at all testable. In other words, a distinction must be made between *fundamental* heuristic hypotheses and *specific* factual assumptions, only the latter being subjected to the correspondence test of observed facts or conditions. The idea of "hypotheses on different levels of generality and, hence, of different degrees of testability" is also emphasized (Hempel, 1965; Machlup, 1978).

But do not such concessions leave the moderate positivist in an untenable position? And is not Popper's position even weaker since he rejects the very thesis of atheoretical observation and yet insists on testing theories against the 'facts'? It seems correct to argue that "if there is no atheoretical observation language then observation statements do not strictly designate the given nor do they strictly express its relational structure". How is it then possible to test the theory against the observation, when the latter does not express the true structure of the real, but is always to some extent theoretical? Only the strict positivist position holding on to the idea of a pure observation language can claim, on the face of it, that the scientist is able to test his theories empirically. But one can still pursue the argument that it is impossible to *demonstrate* that "an observation language does indeed designate the irreducible elements of the real" and that, therefore, "logical positivism is condemned to dogmatism in practice" (Hindess, 1977). Furthermore, the empriricist-positivist model, like any other scheme, is bedevilled by the 'infinite regress' dilemma. Namely, as Popper says, the decision will have to be taken to stop the testing of theories at *some* point, when logically we ought to go on testing the very result obtained by the previous test, *ad infinitum*. Popper appeals here to what he calls the 'inter-subjective' agreement among scientists as to when theories have been sufficiently tested and are regarded as satisfactory, even if only for the time being. He admits that this is a dogmatic position, yet it is innocuous since further testing is always possible if the need arises (Popper, 1961). We hope to add a 'neutralizer' element to the testing procedure by appealing, below, to the everyday world outcomes of theory application. We realize that the outcomes are 'assessments' and these, too, become part of an infinite chain. However, since social theories are not necessarily applied purposely by the scientist, but come to be used in the everyday world often indirectly and by permeation, the dogmatic element in such an 'applied world' is reduced, if not eliminated; certainly insofar as this application of social theories is partly removed from the realm of decision by scientists.

The 'applied science' argument

While it is unlikely that we shall ever be able to extricate ourselves entirely from this epistemological trap, the nihilistic conclusion to which it leads, questioning the very possibility of empirical science is best counter-balanced by the 'applied science' argument. Whether empirical science is dogmatic or not in its foundations, its application to the real world is demonstrably factual, i.e., so far as its results or consequences are concerned. Yet this 'proof of the pudding is in its eating' philosophy, attractive as it is, raises further epistemological and operationalist questions. Thus, if we assume that the application of a theory will in itself alter the course of events, i.e., that scientific results and applications will affect reality, we cannot avoid the question: how can these applications be the cornerstone of testing scientific theories? Simply, does this mean that social reality is so flexible that there is nothing firm against which we can test the theories of social science? Even J.S. Mill, whose classical positivist scheme of scientific enquiry is based on the firm foundation of the regularity and constancy of nature, holds that no such solid base obtains in the domain of the social sciences (Mill, 1967). Karl Popper goes much further and holds that *all* science "does not rest upon rock-bottom". He tells us that there is nothing 'absolute' about objective science, and that: "The bold structure of its theories rises, as it were, above a swamp. It is like a building erected on piles. The piles are driven down from above into the swamp, but not down to any natural or 'given' base; and if we stop driving the piles deeper, it is not because we have reached firm ground. We simply stop when we are satisfied that the piles are firm enough to carry the structure, at least for the time being" (Popper, 1961).

It should be noted, however, that it is one thing to characterize science relativistically because our knowledge and comprehension of the basic nature of our life-world is limited; it is quite another thing to do so because we impute that basic nature with complete plasticity. The evidence does not support such imputation. In the natural sciences the constancy in the numerical relationships between measurable quantities has been accepted in many instances (Machlup, 1978), and even the postulation of uncertainty relationships, involving statistical probabilities, means that physical reality contains certain constant elements. As to the social world, there are those who maintain that social structures are undergoing continuous change, and that the prime sociological task is to explore emerging structures (Glaser and Strauss, 1967). Others point to the seemingly unpatterned aspects of human life, such as the unpredictable chances of war, the ambition and genius of an extraordinary individual, or the fortuitous invention of some new technique (Sprott, 1966). We hold, however, that these factors add new dimensions to the social world, they often jolt the structures, but they do not destroy *au fond* the entire framework of social life. There do exist demonstrable constants in societal life, such as institutional arrangements, interdependence in social groups, social change and exchange mechanisms, etc. There is a certain orderliness, some rhyme and reason, in human life. There is sufficiently clear patterning and firmness to enable us, where our applied knowledge changes the course of events, to find out how and why this has happened. Applied social research will not reach ultimate truth, but it could tell how and how well certain theories work out in practice.

Let us now define 'applied sociology' by adopting a modified version of Paul Lazarsfeld's statement and suggesting that we take it to mean: "Sociological research which is concerned with diagnosing and solving social problems, and which is close to policy formation and practical decision", (Lazarsfeld, 1973). We must at once clarify, though, that there are two sides to the applied sociology coin: that concerned with policy formation, and that which reflects our capacity to test sociological theories and concepts. Our interest here is in the latter. In both cases, however, we must be aware of the complexities and obstacles to be encountered. Thus, many writers have stressed that the complexity of social problems necessitates an interdisciplinary approach, and that narrower analytic approaches of social science disciplines do not fit the wide totalistic problems of the real world situations (Horowitz, 1971, Easton, 1972, Scott and Shore, 1979). Easton believes that to overcome this we must first develop a "philosophy of science in terms of which we seek to validate the whole scientific enterprise" in the social sciences. This would provide the necessary base for the more specialized explanations of each discipline, and enhance the conversion of generalized knowledge into applied form (Easton, 1972, Scott and Shore, 1979).

It is in this vein that in this and previous chapters we sometimes referred to social science research, yet at other times we exemplified our discussion with more specific reference to sociological research. It must be emphasized that we are not using these designations interchangeably. Rather, we see sociological explanation as playing a specific role within the wider framework of social science research, yet bearing in mind that the epistemological issues we have raised are common to both.

The overall problematics of applied research has many ramifications: for instance, contamination or interference effects, investigator-client effects, and the like (Horowitz, 1968, Gouldner and Miller, 1965). Notwithstanding the difficulties, applied research widens the testing ground for sociological theories and it represents a potentially powerful 'natural laboratory'. This potential has, so far, been utilized in a relatively small measure; it has advanced unevenly; and, similarly to pure research, it has been characterized by incompleteness and apparent contradictions. We may derive from our foregoing discussion several possible reasons for such shortcomings: the impossibility of gaining a full knowledge of reality due to its complex nature; or correlatively the limits of our comprehension capacity and potential; and the pluralist nature of scientific techniques arising out of varying interests and contexts. The problematic nature of scientific enquiry to which these possible explanations point need not lead us to the acceptance of anarchism, nor to reject, as Habermas does, the basic methodological unity of all science. Furthermore, incommensurability, which appears to be dominating the scene, is not total and it can be reduced by gaining a greater measure of testability; the latter to be achieved, as we have argued, by the application of sociological knowledge to everday social reality. This provides the real pivot of pragmatic justification for sociology as a science. The 'heuristic' of the pure research programme alone is not sufficient to justify the acceptance of sociological research results and theories. The latter must pass the test of application to real-life situations. Thus, do they help us adapt to and control reality? What are the practical consequences that result from our use of the notions, ideas and principles developed by sociology? Can we successfully plan our actions and organize our behaviour on the basis of sociological knowledge gained from research? In what way do social scientific theories serve the interests of society and its individuals? Such a pragmatist approach (Dewey, 1931, Thayer, 1970, Thayer, 1976) pinpoints the real challenges that confront sociology, and at the same time it provides an invaluable base for testability.

To sum up, we are proposing a flexible model of applied sociological methodology consisting of two interrelated elements: (a) relativistic rationalism; and (b) pragmatic testing. We reach this conclusion because the indeterminism and uncertainty found in the scientific enterprise must, of necessity, lead to a relativistic position. However, this must be wedded to rationalism not in the sense of pure reason, but in the form of critical rationalism, which is tied-in with empiricist science (Prendergast, 1979). For, otherwise, relativism would simply endorse the notion that all research results and ensuing theories are equally true and acceptable (Gellner, 1974, Dean, 1978). Relativism will, it is true, blunt our selector of pragmatic testing, in that it will not reveal a theory to be true or false. The most it can do is to point to a particular theory that works better when applied than another. In this sense no theory is fully validated, and seldom is a theory totally and irrevocably rejected. Similarly no methodology is invalid, but some methodology may be of little use. The basis for social research must be methodological flexibility: namely, we shall adopt that methodological framework which fits best the reality investigated. And the theory which works best, as explanatory and predictive, when applied to solve everyday problems, is the one to be accepted as enlarging our knowledge of reality. The nub of social science is, therefore, pragmatic testing.

Numerous examples can be provided of theories applied with varying 'test' results. The variability of results is to be seen from two angles. First, since we have insisted on the importance of the *pragmatic aspect* of scientific explanation, we shall find that the demands are strictly defined, as Ladislav

Tondl says, "in that the explanation is to make possible sufficiently reliable decisions, dependable predictions and the like". Second, pragmatic application provides a built-in testing mechanism, in that it will show whether a theory explains a phenomenon whenever and wherever encountered; or whether it has a more limited explanatory power. It is in this sense that Tondl mentions the existence of various *levels of satisfactoriness* of explanatory theories, taking into account the possibility of a low level of satisfactoriness where "phenomena can be explained in more than one way or by the use of a greater number of explanatory hypotheses". His more exact definition of 'satisfactoriness' is that: "A scientific explanation may be deemed satisfactory if connected with the possibility of sufficiently dependable prediction, or with that of taking the requisite measures to lead to the fulfilment of the aims intended with the quality intended". In Hempel's formulation this is expressed as follows: "scientific explanation must be potentially predictive and potentially productive" (Tondl, 1973; Popper, 1972).

To illustrate from social research we may take the example of the theory of 'demographic transition' which classifies types of population with different combinations of birth and death rate levels, holding that these 'types' are stages of growth through which a population passes. Thus, the demographic cycle starts with high birth and death rates and a low rate of growth, passes through the stage of high fertility and low mortality accompanied by a very high growth rate, and eventually reaches the low fertility and mortality level resulting in a fairly low rate of growth. This theory, in its simplest form, has not stood the test of demographic trends. Certainly it has not been able to predict population growth patterns in many developing countries; even though it has been useful for the understanding of Western demographic trends (Nam and Gustavus, 1965). A higher level of satisfactoriness has been achieved, however, by sophisticated demographic research showing the influences of ethnicity, education, religion and socio-economic factors on population processes. Thus, the more specific theories arising from such research and stressing the group differences, especially socio-cultural ones, so far as demographic trends in some industrialized countries are concerned, have proved useful for policy objectives, such as the assessment of family planning needs or the projection of population change (Ryder and Westoff, 1971).

Similarly, a fairly high degree of satisfactoriness has been chalked up by sociological theories predicting student unrest and race problems. Of course, as Pettigrew and Back stress, events in the latter sphere "move so swiftly that sociology cannot be reproached for being unable to predict each event as it occurs. Yet this situation argues strongly for an application of a whole range of sociological theory to the study [for instance] of the desegregation process. Many elements of sociological theory, such as relative deprivation considerations, heralded the social crises of the 1960s, but these fruitful leads were not fully exploited". Pettigrew and Back maintain that often there has been a disuse of sociological theory in this area despite its rich promise and relevance and the demonstration of its usefulness (Pettigrew and Back, 1967; Peter Marris, 1964).

Finally, let us consider briefly an example showing both the complex nature of applied social science and the low degree of satisfactoriness sometimes achieved. In the American anti-poverty programme of the 1960s the government agencies applied certain social science theories in an attempt to prevent delinquency in some of the poorest sections of the population. The theories varied and diverged. Some emanated from the work of major sociologists and held, for instance, that there is "an inverse relation between community integration and the rates of juvenile delinquency", hence the maximum feasible participation of the community members themselves was stressed. Another widely held theory claimed that "delinquency arises when socially approved goals .... are made impossible to achieve through legitimate methods". Here 'opportunity' became the master concept. There were other theoretical strands, too, such as the culture conflict one. Low-class delinquency was variously explained as resulting from a search for masculinity of youth reared in female-based households, or as an attempt "to restore the self-esteem that had been savaged by the middle-class dominated institutions". Furthermore, social scientists differed in their views as to the possibility of dealing with delinquency in the manner proposed in some of the anti-poverty programmes. What is worse, the professionals in the field had no confident knowledge of how to prevent delinquency and much of the work was carried out sloppily. There was also rivalry between community action groups, and the wider political and economic circumstances of the 1960s were unfavourable. In some cases one could perceive how "ideological, generational and ethnic problems were compounded by the familiar ones of bureaucracy". Little wonder then that on the whole these programmes failed. But if the blame has to be apportioned to many quarters, Daniel Moynihan does not exclude the social scientists. He questions the extent to which their different theories grasped "a fantastically elusive reality". The social sciences had insufficient knowledge and the evidence was fragmented, contradictory and incomplete. "Why then", he asks, "did the social scientists involved in these events not insist on the limits of their knowledge and methodology?" Moynihan finds the answer in the dual nature of the American social scientist. "He is an objective 'seeker after the truth'. But he is also very likely to be a passionate partisan of social justice and social change to bring it about" (Moynihan, 1969).

Conclusions

The lesson we learn from such real-life experiments is that the social scientist, who in part creates the reality he investigates by defining it and by imparting it with his own brand of scientific meaning, is not free of a controlling hand. His research results and theories must pass a test by being applied to deal with tangible problems. It is this that will uncover for us the efficiency of his theories; their usefulness. It is this that will ultimately show whether, or to what extent, he has improved his understanding of the reality with which he interacts. To uphold such a possibility of pragmatic testability is to call for the results of applied social science to be the final arbiter in the development of sociology.

Bibliography

Albert, Hans, 1974: "The Myth of Total Reason: Dialectical Claims in the Light of Undialectical Criticism" in Giddens, Anthony, ed. *Positivism and Sociology*, Heinemann, London, pp. 158-60, 168-9, 170.

Asher, Herbert B., 1976: *Causal Modelling*, Sage Publications, Beverly Hills, Calif, pp. 12, 50, 69.

Barnes, S.B., 1976: "Natural Rationality: A Neglected Concept in Social Sciences" in *Philosophy of the Social Sciences*, Vol. 6, No. 2. June 1976, p. 115.

Benton, Ted, 1977: *Philosophical Foundations of the Three Sociologies*, Routledge & Kegan Paul, London, pp. 71, 65ff.

Berger, Peter L. & Luckman, Thomas, 1971: *The Social Construction of Reality*. Penguin Books, Harmondsworth, Middlesex, pp. 18, 13, 15, 34, 39. 35ff.

Blackmore, John, 1979: "On the Inverted Use of the Terms 'Realism' and 'Idealism' Among Scientists and Historians of Science" in *The British Journal for the Philosophy of Science*, Vol. 30, No. 2., June 1979, p. 128.

Capek, Milic, 1961: *The Philosophical Impact of Contemporary Physics*, Van Nostrand Reinhold, New York, pp. xvii, 202-3.

Carnap, Rudolf, 1936: "Testability and Meaning" in *Philosophy of Science*, Vol. 3, Baltimore.

Carnap, Rudolf, 1950: *Logical Foundations of Probability*, University of Chicago Press, Chicago.

Carnap, Rudolf, 1956: "The Methodological Character of Theoretical Concepts" in Feigl, H. & Scriven, M., eds., *The Foundations of Science and the Concepts of Psychology and Psychoanalysis*, University of Minnesota Press, Minneapolis, p. 80.

Castell, Alburey, 1976: *An Introduction to Modern Philosophy*, 3rd edition, MacMillan, New York, p. 95.

Coulter, Jeff, 1971: "Decontextualised Meanings: Current Approaches to *Verstehende* Investigations" in *The Sociological Review*, Vol. 19, No. 3, 1971.

Curtis, James E. & Petras, John W. eds.: *The Sociology of Knowledge: A Reader*, Duckworth, London, p. 23f.

Dallmayr, F.R. and McCarthy, T.A. eds., 1977: *Understanding and Social Inquiry*, University of Notre Dame Press, Notre Dame, Indiana.

Dean, John, 1978: "Empiricism and Relativism: A Reappraisal of Two Key Concepts in the Social Sciences", *Philosophy of the Social Sciences*, Vol. 8, 1978, pp. 281-288.

Dewey, John, 1931: *Philosophy and Civilization*, Peter Smith, Gloucester, Mass., pp. 13-35; 318-330.

Douglas, Jack D., 1973: *Introduction to Sociology: Situations and Structures*, The Free Press, New York, p. 97.

Douglas, Mary, 1975: *Implicit Meanings: Essays in Anthropology*, Routledge and Kegan Paul, London, pp. xix, 203-209, 8, 281.

Duncan, Otis Dudley, 1975: *Introduction to Structural Equation Models*, Academic Press, New York, pp. 18-20.

Easton, David, 1972: "Comment on Robert Lane's Political Science and Policy Analysis" in Charlesworth, J.C. ed. *Integration of the Social Sciences Through Policy Analysis*, AAPSS Monograph 14, Philadelphia, pp. 88-98.

Feyerabend, P.K., 1962: "Problems of Microphysics" in Colodny, Robert, G. ed., *Frontiers of Science and Philosophy*, University of Pittsburgh Press, Pittsburgh, p. 195.

Feyerabend, Paul, 1975: *Against Method*, NLB, London.

Garfinkel, Harold, 1967: *Studies in Ethnomethodology*, Prentice-Hall, New Jersey.

Gellner, Ernest, 1973: *Cause and Meaning in the Social Sciences*, Routledge & Kegan Paul, London.

Gellner, Ernest, 1974: *Legitimation of Belief*, Cambridge University Press, Cambridge.

Gellner, Ernest, 1974: "The New Idealism: Cause and Meaning in the Social Sciences" in Giddens, Anthony, ed., *Positivism and Sociology*, Heinemann, London, p. 130f.

Giddens, Anthony, 1974: *Positivism and Sociology*, Heinemann, London, pp. 18-21.

Giddens, Anthony, 1976: *New Rules of Sociological Method*, Hutchinson, London, p. 166.

Gillispie, C.C., 1960: *The Edge of Objectivity: an Essay in the History of Ideas*, Princeton University Press, Princeton, N.J.

Glaser, Barney G. and Strauss, Anselm L., 1967: *The Discovery of Grounded Theory*, Aldine Publishing Company, Chicago, p. 235.

Gorman, Robert A., 1977: *The Dual Vision: Alfred Schutz and the Myth of Phenomenological Social Science*, Routledge & Kegan Paul, London.

Gouldner, A.W. and Miller, S.M. eds.,: *Applied Sociology*, The Free Press, New York.

Habermas, Jurgen, 1970: "Knowledge and Interest" in Emmet D., and MacIntyre, A. eds., *Sociological Theory and Philosophical Analysis*, MacMillan, London, pp. 36-54.

Habermas, Jurgen, 1972: *Knowledge and Human Interests*, Heinemann, London, Part 3, pp. 308-317, 134f., 196f.

Habermas, Jurgen, 1973: "A Postscript to Knowledge and Human Interests", *Philosophy of the Social Sciences*, Vol. 3, pp. 157-89.

Habermas, Jurgen, 1974: *Theory and Practice*, Heinemann, London, Introduction of Chapter 7.

Habermas, Jurgen, 1974: "Rationalism Divided in Two: A Reply to Albert" in Giddens, Anthony, ed., *Positivism and Sociology*, Heinemann, London, pp. 195-223.

Hacking, Ian, 1979: "Imre Lakatos's Philosophy of Science", *The British Journal for the Philosophy of Science*, Vol. 30, No. 4, December 1979, pp. 388, 393, 384.

Heisenberg, Werner, 1930: *The Physical Principles of Quantum Theory*, Dover Publications, New York, pp. 20ff; 62-65.

Heisenberg, Werner, 1958: *The Physicist's Conception of Nature*, Hutchinson & Co., New York.

Heisenberg, Werner, 1975: *Across Frontiers*, Harper Torchbooks, New York.

Hempel, Carl G., 1962: "Explanation in Science and in History" in Colodny, Robert G., ed., *Frontiers of Science and Philosophy*, University of Pittsburgh Press, Pittsburgh, p. 32.

Hempel, Carl Gustav, 1965: *Aspects of Scientific Explanation and other Essays in the Philosophy of Science*, The Free Press, New York, p. 177.

Hindess, Barry, 1977: *Philosophy and Methodology in the Social Sciences*, The Harvester Press, Hassocks, Sussex, pp. 139, 140.

Holzner, Burkhart, 1968: *Reality Construction in Society*, Shenkman, Cambridge, Mass.
Horowitz, Irving Louis, 1968: *Professing Sociology: Studies in the Life Cycle of Social Science*, Aldine Publ. Co., Chicago, p. 228.
Horowitz, Irving Louis, 1971: *The Use and Abuse of Social Science*, Transaction Books, New Brunswick, N.J. p. 3.
Jammer, Max, 1966: *The Conceptual Development of Quantum Mechanics*, McGraw-Hill, New York, pp. 178-180, 167, 168, 173, 330.
Jay, Martin, 1973: *The Dialectical Imagination: A History of the Frankfurt School of the Institute of Social Research 1923-1950*, Little Brown & Co., Boston.
Lachenmeyer, Charles W., 1973: *The Essence of Social Research*, The Free Press, New York, p. 102.
Lakatos, Imre, 1970: "Falsification and the Methodology of Scientific Research Programmes" in *Criticism and the Growth of Knowledge*, ed., by Lakatos, Imre and Musgrave, Alan, Cambridge University Press, Cambridge, p. 180.
Lakatos, Imre, 1978: *The Methodology of Scientific Research Programmes*, Philosophical Papers Vol. 1, ed., by Worrall, John and Currie, Gregory, Cambridge University Press, Cambridge, pp. 6, 68-86, 123, 148.
Lazarsfeld, Paul F., 1973: *Main Trends in Sociology*, George Allen & Unwin, pp. 12, 14.
Lévi-Strauss, Claude, 1976: *Structural Anthropology, Vol. II.* Basic Books, New York, pp. 303-4, 296.
Lévi-Strauss, Claude, 1979: "Science: Forever Incomplete" in *Society* Vol. 16, No. 5, July/August 1979, Transaction, Rutgers The State University, New Brunswick, N.J. p. 18.
Machlup, Fritz, 1978:*Methodology of Economics and Other Social Sciences*, Academic Press, New York, pp. 144f.; 357.
MacRae, Donald, G., 1972: "The Basis of Social Cohesion" in Robson, William A., ed., *Man and the Social Sciences*, London School of Economics and Political Science – George Allen and Unwin, London, p. 58.
Mannheim, Karl, 1936: *Ideology and Utopia*, Harcourt Brace, New York.
Mannheim, Karl, 1952: *Essays on the Sociology of Knowledge*, ed., by Paul Kecskemeti, Routledge, London.
Mannheim, Karl, 1970: "The Sociology of Knowledge" in Curtis, James E., & Petras, John W., eds., *The Sociology of Knowledge: A Reader*, Duckworth, London.
Margenau, Henry, 1950: *The Nature of Physical Reality: A Philosophy of Modern Physics*, McGraw-Hill, New York, p. 377.
Marris, Peter, 1964: *The Experience of Higher Education*, Routledge & Kegan Paul, London, p. 153.
Martin, David, 1973: "The Naked Person: A Critique of Spontaneity" in *Encounter*, Vol. XL. No. 6, June 1973, p. 12.
McCarthy, Thomas, 1978: *The Critical Theory of Jurgen Habermas*, Hutchinson, London, pp. 411-12, 49, 45, 299f., 353.
McLachlan, Hugh V., 1980: "Popper, Marxism and the Nature of Social Laws", *British Journal of Sociology*, Vol. 31, No. 1. March 1980.
McSweeney, Bill, 1973: "Meaning, Context and Situation" in *Archives Eurpeennes de Sociologie*, Vol. XIV, No. 1, pp. 143, 148-9.
Meinong, Alexius, 1902: "Über Annahmen", *Zeitschrift für Psychologie und Physiologie der Sinnesorgane*, Supplementary Vol. II, Leipsic. Quoted in "Realism" by Joad, C.E.M., *Encyclopaedia of the Social Sciences*, ed., Seligman, E.R.A., vol. 13, MacMillan, New York, 1950.
Merton, Robert K., 1957: *Social Theory and Social Structure*, The Free Press, New York, pp. 469-471.
Mill, J.S., 1967: *A System of Logic*, Longman, London, p. 554.

Moynihan, Daniel P., 1969: *Maximum Feasible Misunderstanding*, The Free Press, New York, pp. 50-53, 75f., 106f., 128, 139, 167f., 170f., 173-180, 188-197.

Nam, Charles B. and Gustavus, Susan O., 1965: *Population: The Dynamics of Demographic Change*, Houghton Mifflin Co., Boston, pp. 36-39.

Overend, Tronn, 1978: "Enquiry and Ideology: Habermas' Trichotomous Conception of Science". *Philosophy of the Social Sciences*, Vol. 8, No. 1, p. 8.

Pettigrew, Thomas F. and Back, Kurt W., 1967: "Sociology in the Desegregation Process: Its Use and Disuse" in P.F. Lazarsfeld, et al., eds., *The Uses of Sociology*, Basic Books, New York, p. 717.

Popper, Karl R., 1961: *The Logic of Scientific Discovery*, Science Editions, Inc., New York, pp. 104-5; 111.

Popper, K., 1963: *Conjectures and Refutations*. Routledge and Kegan Paul, London.

Popper, Karl, 1970: "The Sociology of Knowledge" in Curtis, James E. & Petras, John W., eds., *The Sociology of Knowledge: A Reader*, Duckworth, London, pp. 652, 653.

Popper, Karl R., 1972: *Objective Knowledge: An Evolutionary Approach*, The Clarendon Press, Oxford, pp. 158ff., 193, 328f., 329, 304.

Popper, Karl R., 1973: "Indeterminism is Not Enough" in *Encounter*, Vol. XL. No. 4, April 1973, p. 26.

Popper, K., 1974: *The Philosophy of Karl Popper*, P.A. Schilpp, ed., The Library of Living Philosophers, Vol. XIV, The Open Court Publishing Co., Illinois, p. 980f.

Prendergast, Christopher, 1979: "The Problem of the Unity of Knowledge in Comte's Philosophy of Science", *Sociological Inquiry*, Vol. 49, No. 4. pp. 25, 26.

Quine, W.V.O., 1960: *Word and Object*, MIT Press, Cambridge, Mass., p. 76.

Robertson, Roland, 1972: "The Sociocultural Implications of Sociology: A Reconnaissance" in T.J. Nossiter et al., eds., *Imagination and Precision in the Social Sciences*, Faber & Faber, London, pp. 77, 90f.

Rosenberg, Alexander, 1974: "Partial Interpretation and Micro-Economics" in Leinfeller, W. & Kohler E., eds., *Developments in the Methodology of Social Science*, D. Reidel Publ. Co., Dordrecht, Holland, pp. 94-109.

Ryder, Norman B., and Westoff, Charles F., 1971: *Reproduction in the United States 1965*, Princeton University Press, Princeton, N.J. pp. 53, 265.

Scheler, Max, 1970: "On the Positivistic Philosophy of the History of Knowledge and its Law of Three Stages" in Curtis, James E., and Petras, John W., eds., *The Sociology of Knowledge: A Reader*, Duckworth, London, pp. 165, 175.

Schilpp, Paul Arthur, 1963: *The Philosophy of Rudolf Carnap*, Open Court, La Salle, Ill. p. 59.

Schutz, Alfred, 1972: *The Phenomenology of the Social World*, MacMillan, London, p. 61.

Scott, Robert A. and Shore, Arnold R., 1979: *Why Sociology Does Not Apply*, Elsevier, New York, p. 225.

Simon, Herbert A., 1969: *The Sciences of the Artificial*, The M.I.T. Press, Cambridge, Mass.

Sprott, W.J.H., 1966: *Sociology*, Hutchinson, London, p. 56.

Stark, W., 1958: *The Sociology of Knowledge*, Routledge & Kegan Paul, London, p. 169.

Thayer, H.S. ed., 1970: *Pragmatism: The Classic Writings*, New American Library, New York.

Thayer, H.S., 1976: "Pragmatism", in *Encyclopaedia Britannica*, Macropaedia, Vol. 14, Chicago, pp. 940-944.

Tilley, Nicholas, 1980: "Popper, Positivism and Ethnomethodology", *British Journal of Sociology*, Vol. 31, No. 1, March, 1980, p. 31.

Tondl, Ladislav, 1973: *Scientific Procedures*, Boston Studies in the Philosophy of Science, Vol. X, D. Reidel, Dordrecht, Holland/Boston U.S.A., pp. 155-6, 179.

Weber, Max, 1972: *The Interpretation of Social Reality*, ed., by Eldridge, J.E.T., Nelson, London, p. 12.

Wolff, Kurt H., 1970: "The Sociology of Knowledge and Sociological Theory" in Curtis, James E., and Petras, John W., eds., *The Sociology of Knowledge: A Reader*, Duckworth, London, pp. 546-7.

Znaniecki, Florian, 1963: *Cultural Sciences*, University of Illinois Press, Urbana, p. 11.

Chapter 6

Conclusion

The foregoing discussion of the principles of social scientific methodology has, especially in the last chapter, brought us close to the central questions concerning the basic nature of social life and the epistemology most suitable for it. It will be useful to present in these final remarks, briefly, the crucial issues concerning the above questions, and to put forward our own viewpoint concerning these issues.

Two extreme views can be discerned easily. At one end we have the negative standpoint stressing the non-generalizability of social events, seeing society as continually changing and emergent, and being decidedly opposed to a physicalist model for the social sciences. We can trace back this line of thought from latter-day phenomenologists and Wittgensteinian contextualists, like A.R. Louch and Peter Winch, to a more mitigated Humean scepticism. Such scepticism, however, could be simply instrumental in counteracting the other extreme, represented by the unbridled faith of a Bridgman or a Lundberg in operationalism, quantification, and a determinist scientific model, still believed in by a powerful band of social scientists. In stressing fallibilism and the permanent possibility of falsity in even the best confirmed empirical generalizations and predictions, Humean scepticism paved the way for Popperian philosophy, thus replacing inductive inconclusiveness, not by deductive certainty but by a provisional confidence based on whatever empirical demonstration and testing is attainable (Hume, 1955; Stove, 1973; Penelhum, 1975; Gellner, 1979).

It is commonly accepted now that empirical science has its limitations. But these exist not because the physical world is basically inscrutable. Its regularities and constancies are there to see in the everyday world, yet the explanations scientists offer us at any moment are partial explanations for, as Bronowski says, the world is totally connected but science can grapple only with partial connections (Bronowski, 1978). It appears that the scientist, who has to carry out a 'decoupling' process before he restores the particular connections necessary to see a slice of reality, gains an imperfect understanding of even that specific isolated part of reality despite the regular and stable aspects he uncovers. The social world, as the physical world, has its stable dimensions, its constant elements, side by side with its dynamism reflected in the different types of change, ranging from smaller fluctuations to developmental and more basic changes. But its units, its entities, its concepts, be they the social system, the primary group, or social class, are no less and no more real than the units, entities and concepts of physics, whether mass, energy or electron.[1] And whilst true that the research apparatus used in sociology cannot be the same as that used in physics, the underlying methodological principles are common to them both.

These principles serve as guidelines for our scientific attempts to uncover the complex regularities of the world and to offer causal explanations of events. But there is disagreement among scientists and philosophers of science on whether the methodological principles rest on a deterministic or indeterministic world-view and, therefore, whether science can ultimately offer us only fragmentary knowledge and probabilistic predictions, or perhaps arrive at much firmer foundations of knowledge. This is obviously why the scientist can be characterized in the way P.B. Medawar sees him, as someone with "a certain mixture of confidence and restless dissatisfaction" (Medawar, 1967). The dissatisfaction is due, we think, to the uncertainties the scientist must always face. This can lead, however, to greater excitement in scientific work, and since this volume is about the methodological problems of the sociologist, we fully concur with Clark Glymours's view that: "For a methodologist, the social sciences today are in some ways more exciting than is contemporary work in physical science. Exactly in those fields where impressive and dominant results are difficult to obtain, methodological considerations are likely to be most explicit, and innovations in method are likely to occur most often. So it is in contemporary work in a field like sociology" (Glymour, 1980). One consequence of applying the general ongoing debate in the philosophy of science to social science methodology is the impetus that this gives to constant comparisons between the physical and social sciences and the resultant attempt to relate certain explanatory principles "of the easily measured features of the world to less easily measured features and where consequently there is much less *a priori* belief as to what kinds of hypotheses should explain what kinds of empirical regularities" (Glymour, 1980). The possibility of applying such principles to sociology, especially the principles of probabilistic explanation and prediction, which is a feature of all science, brings social research into the scientific fold.

We ought to mention here the two aspects of probability: the statistical meaning which refers to the objective regularities discovered, and the logical meaning which refers to the strength of our belief in scientific findings. Both these dimensions highlight the dual character of probabilistic explanation

[1] The question has been raised whether the physicist has actually *seen* the subatomic particle called neutrino when he points to scratches on photographic negatives which he claims to be its tracks (Gregory, 1981).

and prediction, namely the contingency of the given relations and the basically uncertain character of our prediction, as well as the relatively stable tendency of, say, the occurrence of X in relation to Y. That is to say, the results of our research in the social sciences can describe, explain, or predict, only 'conditional probabilistic relations'. Our predictive premises consist of random relations in the weak sense, so that where we have "numerous 'natural' clusters or 'sequences' ", we attain statistical randomness not of the 'perfect coin toss' variety, but one that will be affected by a large number of *potential modifiers* (Nowak, 1976). Such is social reality. Attempts will be made, therefore, to stabilize or strengthen the predictions and explanations by relating them to particular historical and geographical populations or cultural contexts. At any rate, we simply cannot disregard the conditional character of any such prediction or explanation. Therefore, the degree of predictive power an explanatory theory possesses depends on what empirical information is available concerning existing conditions and potential modifiers to these conditions. One thing is certain: classical determinism with its model of 'the perfect predicter' does not apply here. In its place comes scientific thinking along the lines of "the probability of possible future states" and degrees of predictive accuracy (Nowak, 1976).

Putting sociological research into a methodological framework which relies on statistical probability statements means that such work can be regarded as 'scientific' in the wider indeterminist sense. This does not, however, eliminate the major problem that such probability statements in sociology often lead to such weak forms of explanation as to render them not much more valuable than intelligent guesswork. This lack of scientific success is usually explained by the practical inability of the social researcher to engage in scientific experimentation within the framework of sociological methodology. Another argument is that the lack of success is due to the mistake that many social researchers make in trying to copy the scientific model of the physical sciences.

Although it is true that scientific experimentation along the lines of the natural science model cannot be applied in social research, sociology *can* substitute for the artificial experimentation of the physical sciences its own brand of experiment. This takes place in the real life-world which is sociology's natural testing ground for its theories and findings. In the long-run, as in all the sciences, it is within this applied framework that the discipline will either find supporting evidence for its theories or, alternatively, find them to be discredited and thus falsified. The fact that the hope lies in the applied field must not lead, however, to the illusion that all the problems of sociological research will ultimately be solved, that in the end the absolute will be achieved, that apodictic knowledge is within our grasp. On the contrary, we have shown how numerous are the *unresolved problems*, and have highlighted the complexities and uncertainties of even accurate sociological description, let alone causal explanation and prediction. But this is the nature of indeterminist science: not to regard the world as inscrutable, but to seek explanation by pushing against the limits, by overcoming obstacles, by extending the range, by widening the horizon of knowledge; yet settling for what appears to be correct even if it is bound to remain incomplete (Jammer, 1979). Sociology, with its methodological postulates, cannot hope for more than this.

Although we must acknowledge the distinctive character of sociology, with its human rather than material subject matter and, therefore, its inability to simulate experimentally the reality it wishes to probe, it is true to say that some of the deeper problems met in sociological research are common to all scientific disciplines. One such basic problem is that, in the last resort, human judgement is involved in all scientific work. As an example, how correct is our judgement that 'relevance' has been achieved in the investigation of any aspect of reality, physical or social? In this respect we inevitably face the insoluble dilemma of how we can know that our awareness of some datum represents 'the immediately given'. It seems to be true, therefore, that the assertion that "empirical knowledge is confined to what we actually observe" is a tautology (Schlick, 1949). The seeking of pure empiricism will not in itself conjure away such a basic problem. Indeed, social scientists do not delude themselves into thinking that they can achieve complete relevance. It is always acknowledged that even slices of social reality cannot be fully reconstructed, that knowledge of social reality has never absolute empirical grounding, that we are never certain against what ultimate social reality we are setting our empirical research; and that to some degree social research is always value-relevant, so that even the synthesis arrived at by social science can only achieve an unpure and relative, rather than definitive, statement regarding the phenomena of the social world (Mannheim, 1960, Lessnoff, 1974). It might seem inconsistent to claim that sociology is a science, yet that it cannot ultimately know what it deals with! Perplexing as this might appear to be, it must be acknowledged that partly such inconsistency prevails; but not only in sociology. Had ultimate reality been known in the physical sciences and in the other social sciences, many universal mysteries and human miseries would have disappeared overnight. But these perplexities of the physical and the social world must not deter the scientist from making bold assumptions and from proceeding methodically to accumulate knowledge. Neither must they deter the sociologist from undertaking bold research based on methodological principles of controlled investigation and practical testing. Although the resulting explanations and predictions will remain probabilities rather than absolute certainties, the sociologist can make a valuable contribution to science and society.

Bibliography

Bridgman, P.W., 1928: *The Logic of Modern Physics*, Macmillan, New York.

Bronowski, Jacob, 1978: *The Origins of Knowledge and Imagination*, Yale University Press, New Haven, pp. 58, 96, 100.

Gellner, Ernest, 1979: *Spectacles and Predicaments: Essays in Social Theory*, Cambridge University Press, Cambridge, Chapter 3.

Glymour, Clark, 1980: *Theory and Evidence*, Princeton University Press, Princeton, New Jersey, p. 291.

Gregory, Richard L., 1981: *Mind in Science*, Weidenfeld and Nicolson, London; reviewed by John Naughton in *The Observer*, London, 3rd of January, 1982.

Hume, David, 1955: *Enquiry Concerning the Human Understanding*, ed. by C.W. Hendel, Bobbs-Merrill, New York.

Jammer, Max, 1978: "Physics and the Search for the Absolute" in *The Search for Absolute Value in a Changing World*, Proceedings of the 6th International Conference on the Unity of the Sciences, International Cultural Foundation Press, New York.

Lessnoff, Michael, 1974: *The Structure of Social Science*, Allen and Unwin, London, pp. 163-5.

Louch, A.R., 1966: *Explanation and Human Action*, Basil Blackwell, Oxford.

Lundberg, G.A., 1939: *Foundations of Sociology*, Macmillan, New York.

Mannheim, K., 1960: *Ideology and Utopia*, Routledge and Kegan Paul, London, pp. 26, 51.

Medawar, P.B., 1967: *The Art of the Soluble*, Methuen, London, p. 110.

Nowak, Stefan, 1976: *Understanding and Prediction: Essays in the Methodology of Social and Behavioral Theories*, Reidel, Dordrecht, Holland/Boston, U.S.A., pp. 231-241.

Penelhum, Terence, 1975: *Hume*, The Macmillan Press, London, pp. 26-35, Chapter 2.

Schlick, Moritz, 1949: "Meaning and Verification" in Herbert Feigl and Wilfrid Sellars, eds., *Readings in Philosophical Analysis*, Appleton-Century-Crofts, New York, p. 152.

Stove, D.C., 1973: *Probability and Hume's Inductive Scepticism*, Clarendon Press, Oxford.

Winch, Peter, 1958: *The Idea of a Social Science*, Routledge and Kegan Paul, London.

Name index

Subject index